The
Aspen
Institute

Change and Prosperity:

The Aspen Institute Program on the World Economy

E. Gerald Corrigan and William D. Eberle, Co-Chairmen
Joan E. Lovett, Director

Report of the Conference
August 17–21, 1996
Aspen, Colorado

The Aspen Institute
Suite 1070
1333 New Hampshire Avenue, N.W.
Washington, D.C. 20036
Published in the United States of America
in 1997 by The Aspen Institute

Copyright © 1997 by
The Aspen Institute

All rights reserved

Printed in the United States of America

ISBN # 0-89843-213-8

Table of Contents

Acknowledgements 5

1996 Conference Report. 7

Commissioned Papers
*"Reflections on Globalization, Technological Change,
and the Labor Market"*
      Lawrence F. Katz 21
*"Unemployment and Wages in Europe and
North America"*
      Stephen Nickell. 51

Appendices 93

List of Participants 93

Conference Program. 97

Questions for Discussion 101

Acknowledgements

The objective of "Change and Prosperity: The Aspen Institute Program on the World Economy" is to help forge an international consensus on possible solutions to critical problems facing the world economy. Toward that end we annually convene a diverse group of business and financial leaders, senior-level government officials, legislators, academic experts, and media representatives – 39 leaders from 12 countries in August 1996—to discuss a carefully focused agenda. We are grateful to our fellow participants, whom we list at the end of this conference summary, for sharing with us their rich experience and insights. The summary of the deliberations at the 1996 Conference reflects a synthesis developed by the Co-Chairs and is not attributable to individual participants.

Our conference was made possible by the generous support of the following organizations and individuals. "Change and Prosperity: The Aspen Institute Program on the World Economy" and The Aspen Institute gratefully acknowledge their contribution.

BankAmerica Foundation
Bankers Trust Company
    Foundation
Center for Global Partnership
Centre for Strategic and
    International Studies
Citibank, N.A.
Dow Jones Foundation
International Monetary Fund

Goldman, Sachs & Co.
Japan Economic Foundation
Robert McNamara
VEBA
Wanger Asset
    Management, L.P.
Warburg, Pincus Ventures, Inc.
The World Bank

We also want to acknowledge the extraordinary role played by John V. Moller in contributing to the success of the Program over the years. For personal reasons, Mr. Moller has relinquished his position as Director of the Program. Mr. Moller has been succeeded by Ms. Joan E. Lovett, who had been a Senior Vice President at the Federal Reserve Bank of New York.

E. Gerald Corrigan
William D. Eberle
April, 1997

1996 Conference Report

Introduction

We live in a time of momentous and continuous change, driven by a revolution in technology and an increasingly integrated international economy and global financial system. The state of the world economy today is reasonably sound and the macroeconomic outlook for the near term is broadly satisfactory although major problems loom over the longer term. Market-based economic policies and democratic institutions have taken root throughout much of the world. The community of nations, as a consequence, finds itself on the threshold of an opportunity to create an environment in which broad-based economic progress can be achieved on a scope and scale of virtually unprecedented proportions.

Periods of great change are invariably also periods of instability and anxiety, however. That is certainly true today. People in many countries are growing impatient to see the fruits of "sound policies" reflected tangibly in their daily lives. At the same time, through the revolution in global communications, awareness of growing disparities in wealth and income—the gap between "haves" and "have nots"—within countries and between countries is increasing. Instability and anxiety can breed discontent and threaten countries' pursuit of "sound policies" and their commitment to economic and political reform.

There is, therefore, no guarantee that we will seize the opportunity before us and reap the benefits it holds. Our future depends on the wisdom of the policy choices we make as individual nations and as a global community. Defining some of those key choices and pointing the way forward was the purpose of our conference.

Macroeconomic Policies for Growth and Adjustment

The state of the world economy today is reasonably sound, owing mainly to the continuing strong performances by the United States, the U.K., and a cohort of emerging market countries. Looked at from the vantage point of the closing months of 1996, the macroeconomic outlook in the period immediately ahead is reasonably bright. No urgent problems threaten. No major crises loom.

Over the medium and long-term, however, we face a broad array of formidable challenges. Four in particular were the focus of our discussions.

First, growth prospects in Japan and much of Europe are, at best, uncertain. Japan's economy is stirring after four years of stagnation but it is uncertain if the resumption of growth in domestic demand in Japan can be sustained in the absence of continued government stimulus measures that would exacerbate a large budget deficit. Fear that Japan's recovery may falter is dampening investor confidence out of concern that Japan's huge overhang of bad real estate debts will resurface, creating the risk of a vicious circle.

Europe has experienced seven years of sluggish growth. Nevertheless, Europe is determinedly embarked upon a highly ambitious path toward monetary union, the creation of a single European currency and central bank by 1999. Achieving this goal will require many European governments to move toward further fiscal consolidation which will hamper prospects for growth for at least a period of time.

Second, the aging of populations of the industrial countries portends an explosion of demands for retirement and health care benefits less than a decade away. The unfunded pension liabilities facing industrial country governments are enormous. They hold far-reaching implications for efforts to manage fiscal deficits and raise troubling inter-generational equity issues. They demand that difficult decisions to better rationalize the structure and costs of deeply embedded social welfare programs be faced now.

Third, the incidence of very high rates of unemployment in Europe, stagnant or falling real wages for important segments of the work force in the United States, and sharply skewed patterns of income distribution in almost all emerging market and transition economies raise serious questions for the longer term about social cohesion, economic and political stability.

Fourth, The domestic financial systems of many emerging market countries are fragile and they are vulnerable to macroeconomic shocks. Mexico's experience taught us all a hard lesson. Given the risks of contagion of national economic crises, the strong emphasis given by the G–7 countries in their 1996 communiqué at Lyon to the international financial system and the institutions that manage and participate in it, was highly appropriate. Their call for more effective supervision of financial markets and institutions in emerging market countries is appropriate. However, improved supervision itself is a necessary, but not remotely sufficient condition to building sound, effective and efficient national banking and financial systems in many emerging market and transition countries.

A key challenge on which we focused particular attention was that of managing the process of the full integration of the emerging market countries into the global economy. Their ability to grow and raise the living standards of their, in many cases, rapidly growing populations will depend upon their ability to gain and maintain access to the goods, services and investment capital of the industrialized nations. Likewise, the ability of the industrial countries, with their mature, slower-growing economies, to maintain balanced growth depends importantly on the availability of export markets for their high value added products and services and a stable economic and political environment into which they can channel cross border investment flows.

In these circumstances, patterns of more or less synchronized growth in the global economy presupposes a still higher and deeper degree of integration of the economic, financial, and institutional ties between nations. Of particular importance in this regard is greater integration between the industrial countries, with their mature institutions, and the emerging market countries where, institutional reform and institutional integration represent huge challenges.

Integrating Emerging Market Countries into the Global Economy: Latin America as a Case Study

While the great diversity of Latin American countries makes generalizations difficult, it is fair to say that, on the whole, over the past 10 to 15 years, Latin America has made sweeping changes—embracing market-based economic policies and democratic institutions—and achieved considerable progress in macroeconomic terms. In certain areas, notably fiscal policy, the performance of Latin economies compares favorably with that of the G-7 and even, surprisingly, the emerging market countries of East Asia. Yet, for a variety of reasons including low domestic savings rates, Latin America has failed to achieve vigorous rates of growth on a sustained basis. Further, income growth per capita has been negligible, and many people have failed to see macroeconomic achievements reflected in improvements in their daily lives. Against this background we addressed the question of what the countries of the region have to do to establish a durable foundation for sustained, vigorous rates of economic growth and improved living standards for all their people.

The experience of countries that have successfully made the transition from putting into place the requisite sound policies to reaping benefits in terms of higher living standards indicates that the process is a long, arduous one requiring politically difficult decisions. It demands perseverance in the application of consistent, disciplined policies over an extended period of time. Consistency is especially important in establishing predictability, and predictability is the key to establishing credibility for a country's economic policy managers, an essential prerequisite for full integration into the global economy.

For many countries—including most of the Latin American nations—the challenge in the period ahead is particularly difficult in that coming full circle in the reform and integration process will require concentrated effort on a number of fronts in a setting in which the burdens of reform and adjustment are heavy, especially for the lower income segments of the population.

The demands associated with the advanced stages of that transition process can be mitigated and the process could be accelerated by a concentrated and multi-dimensional policy framework that includes the following elements:

- Macroeconomic Policy—fiscal policy, including tax reform and monetary policy, including exchange rate policy.
- Structural Policies—initiatives, especially pension reform, designed to increase domestic savings, as well as infrastructure development, trade and foreign investment policies, and labor market policies.
- Institutional Reform—policies that more deeply establish the rule of law and framework for the operation of a market-based economy—judicial, legal, and regulatory.
- Financial Reform—reforming the structure of banking systems, instituting effective regulatory and supervisory systems, and developing a credit culture through which the rights and obligations of both debtors and creditors are more universally understood and applied.
- Official Sector Financial Management—developing sophisticated approaches to internal and external debt management, interest rate, exchange rate, and liquidity risk management in order to better guard against internal or external financial shocks.

With these policy mandates in mind, the group discussion focused on the critical issue of maintaining public support for the reform process. There was considerable disagreement within the group in this regard. Some argued that the process of reform should be accelerated while others, especially participants from the Latin American countries, favored a more deliberate approach, in part because they believed that the capacity to absorb changes was already stretched to the limit.

While there were differences of opinion as to the pace of reform, there was complete agreement that financial markets would inflict a very harsh verdict on any country in which there were serious shortcomings in policy and performance.

Finally, given all the difficulties and the perils involved, the discussion turned to the question of whether some countries might choose to "opt out" of the global system in sole pursuit of an internal agenda. It was recognized that the potential costs of opting out to an individual country were so large that few, if any, countries will choose that approach. But, it was also agreed that even if a large country chose to opt out, the consequences for the balance of the world economy would be slight.

A World Bank study confirms this. It compared the economic performance of the 25 developing countries most and least integrated into the global economy. The results were unambiguous: the 25 most open and most integrated countries experienced income growth 3 percent faster than the 25 least integrated. That 3 percent means the difference between doubling incomes in a generation and stagnant incomes, demonstrating that emerging market countries have no choice but to integrate into the global economy.

Four major issues emerged as recurring themes through our continuing discussions.

1. **Inflation.** While there was universal agreement that excessively high (more than 25 percent) rates of inflation are inherently unstable and that low inflation promotes economic growth and stability, there were sharply differing views among some participants as to the speed with which high inflation rates should be reduced. The debate of "shock treatment" versus "gradualism" was inconclusive.

2. **Exchange Rates.** Overvalued exchange rates, it was generally agreed, posed a risk of precipitating balance of payments crises. That being the case, it was also agreed that for many countries, greater attention needs to be paid to the determinants of and monitoring of real exchange rates.

3. **Managing Capital Flows.** There was a strong consensus that most emerging market countries should adopt a more aggressive posture with regard to managing elements of both current and capital accounts. As a part of this, internal and external debt management policies need to be strengthened. Similarly, more sophisticated techniques to assess interest rate, exchange rate, and liquidity risks are also needed. Within this context, there was a lively debate as to

whether such efforts to better manage capital flows should include selective administrative actions such as have been successfully employed by, for example, Chile. While it was acknowledged that administrative efforts to protect against speculative short term capital flows may be useful in certain situations, there was great concern that such initiatives can all too easily become the first steps down the slippery slope leading to undesirable and ultimately counterproductive capital controls.

4. **Unemployment.** The group saw little likelihood that significant reductions in unemployment levels in most Latin American countries could be achieved in the near term. This problem is amplified by the fact that for many displaced workers—to say nothing of new labor market entrants—it will take years before such workers can be fully assimilated into the labor force. All of this underscored to the group the critical importance of structural policy initiatives aimed at labor market reform, worker training, and education.

To summarize, all of the program participants acknowledged that great strides are being made in virtually all countries in Latin America. Of particular note, was the relatively rapid recovery in Mexico and elsewhere following the peso crisis of 1995. Indeed, while the challenges that lie ahead are formidable, the discussion concluded on an upbeat note reflecting in large part the continuing commitment to sound policies on the part of the economic and political leadership in the region.

The Socio-Economics of Growth and Income Distribution— A Global Perspective

This is, in familiar terms, the challenge of narrowing the gap between "haves" and "have nots." A fundamental challenge facing all societies is how to ensure that the distribution of the benefits of growth serves as an incentive to produce more growth while at the same time promoting greater social justice and values supportive of market economics and pluralistic politics. Over the long term, failure in meeting this challenge portends serious social, economic, and political consequences that could undermine public support for sound economic policies and democratic institutions.

We focused our discussion on the employment problems being experienced by the industrial countries because, increasingly, those problems—high rates of structural long-term unemployment in Europe, real wage stagnation, and growing income inequality in the United States—are being attributed by workers and the politicians eager to attract the support of those workers, to the process of global economic integration.

To examine that proposition, we commissioned two papers, one from an American perspective and one from a European perspective[1], diagnosing the employment maladies of the U.S. and the EU, to serve as a foundation for our discussion of the policy implications. Drawing heavily on the two papers, a general consensus emerged on several critical points as follows:

First, while income inequality in much of Europe has not increased as rapidly as in the United States, patterns of employment growth and unemployment rates are much worse in Europe than in the United States. This striking difference was largely attributed to more flexible labor markets in the U.S. and broader and deeper social safety nets—with correspondingly high rates of government spending relative to GDP—in Europe. However, whether the problem was growing income inequality or growing unemployment, the results were seen as distinctly troublesome.

Second, greater global interdependence—particularly imports of manufactured goods into industrialized countries from "low wage" developing countries—was not seen as the major factor explaining labor market developments in either Europe or the United States. Relative to GDP, such imports are simply too small to have such a large macroeconomic effect. This is not to say that this phenomena is not a factor—especially in some sectors of the economy— but rather that it is not as important as suggested by some. On the other hand, some participants suggested that patterns of population migration were aggravating the situation in economic and political terms.

Third, the single most important factor helping to explain falling real wages in the U.S. has been the major change in skill-based technology as it manifests itself in changing patterns of demand in the labor market in a setting in which there has been a failure of education and training programs to respond more effectively to the changing skill requirements of the market place.

Fourth, it appears that more effective education in Europe does help to explain why European workers seem to be more adept at coping with shifts in demand in favor of higher skilled jobs even if questions remain as to how all of this relates to the unemployment problem.

Finally, since neither the trends toward more global integration nor changes in skill based technology have run their course, there is great uncertainty as to how these trends will reflect themselves in the years ahead. On the trade side, for example, the potential weight of China and India are major imponderabilities. One thing, however, is clear; in confronting these trends, the past will be a very imperfect guide to the future.

Against this background, the discussion turned to the policy implications of these observations. In discussing policy options, there was universal agreement that sustained growth was a necessary, but not sufficient condition to solve the income inequality and high unemployment problems confronting much of the industrial world. That is, structural policies in regard to training and education as well as basic reforms in welfare and related social safety net design will be needed. Within this context, it was recognized that policy initiatives had to focus on two broad categories of the population, as follows:

First; prospective new labor market entrants with special emphasis on the school-aged segment of the population. Here it was recognized that stronger basic education was critical but also that special programs targeted at the disadvantaged were equally important. In this regard, a number of participants cited examples of highly successful programs in some communities in such areas as (1) school-to-work transition for non-college bound students; (2) educational, health care and child care programs for school-aged un-wed mothers; and (3) pragmatic job oriented computer training programs for high school students. After listening to the description of these "success stories", the group was both encouraged that such programs are working and mystified as to why they are not more widely in use.

Second; older workers who have been displaced by technology and are now either unemployed or working at significantly lower wages than earlier in their careers. For these workers, the approaches needed are quite different than is the case for young school aged people in that re-training is the key. In discussing this aspect of the problem, the group noted that private sector employers have a heavy responsibility to bear in coping with these problems. Greater attention, for example, needs to be paid by employers to (1) continuing education and training in the work-place; (2) corporate sponsored retraining programs; and (3) corporate sponsored transition and job search programs for displaced workers.

While targeted programs sponsored at the corporate or community level have great potential and attraction, it was also agreed that basic reforms in education and welfare systems were needed with somewhat greater emphasis on the education system in the U.S. and on the welfare (or social safety net) system in Europe. There was also a surprising degree of support for governmentally supported supplements for low wage jobs for at least some period of time. In this regard, participants had in mind devices such as "in work benefits" in Europe and the earned income tax credit for low income individuals in the U.S.

The group concluded its discussion of this subject on a note of concern growing out of the widely accepted view that even under the best of circumstances it will take considerable time, effort, and money to effectively deal with these issues. In these circumstances it was stressed that leaders in both the public and private sectors have a major role to play in building public support for the necessary reforms and programs.

The Politics of Economic Policymaking

This discussion focused on how better to insure that governments and parliaments make and sustain sensible policies in the face of potent short-term political pressures to act in ways that are inconsistent with long-term needs. In other words, how can governments build and sustain the political will that is essential to embrace a disciplined intermediate and longer-term policy orientation in the face of election cycles and other shorter-term pressures and demands.

A serious obstacle, we concluded, is the erosion of trust in government leadership and institutions. This was a theme that pervaded our discussions, a nearly universal phenomenon in industrial and emerging market countries alike. Governments and governmental institutions have lost peoples' confidence because, in part, of what one participant characterized as "the iron law of democracies"—that if addressing a serious problem can be deferred, it will—and in part because the theatrics of politics have undermined the environment for good policy making. Building public trust must be a top priority for governments everywhere.

Part of the solution lies in the importance of government "marketing"—i.e., devoting the required time and effort to explaining and justifying difficult policy decisions. This task is especially important in countries that have newly made the transition from authoritarian to democratic political systems though it is important in all countries.

Responsible leadership is needed not only at the national level but also internationally. Strong domestic political pressures in the United States are pushing in the direction of a lessening of the traditional leadership role of the U.S. in promoting open market based economic systems and democratic institutions on a global scale. Europe is too preoccupied, obsessed some said, with the challenges of its own integration to become a full-time partner of the United States in helping manage the emerging global economic order and Japan is weighed down by a complex and demanding domestic agenda. On the other hand, the group recognized that all nations must, inevitably, place the domestic economic agenda first in a setting in which the I.M.F. and institutions such as the G-7 can play a constructive role in helping countries do what they must do in policy terms in any event. Taking the longer view, however, achieving the great promise of global prosperity will surely require greater and more effective cooperation as well as financial and moral support for the multinational institutions including the World Bank, the I.M.F., and the W.T.O.

We concluded that discussion by identifying what we considered to be two additional essential tasks for the future:

1. Reforming our international institutional architecture to integrate the key dozen or fifteen leading emerging market countries into the process of global governance and international decision making.

2. Preventing the marginalization of the least developed countries whose disintegration holds substantial risks in environmental and other, if not economic, terms for the rest of the world community.

As the Conference concluded, it was agreed that the major theme for the 1997 meeting would focus on the task of achieving a higher degree of global integration in economic, financial and political terms.

End Notes

[1] Katz, Lawrence F. "Reflections on Globalization, Technological Change, and the Labor Market." Nickell, Stephen "Unemployment and Wages in Europe and North America."

Reflections on Globalization, Technological Change and the Labor Market

Lawrence F. Katz

Harvard University and National Bureau of Economic Research
August 1996
Revised May 1997

This paper was prepared for a meeting of The Aspen Institute Program on the World Economy, Aspen, Colorado, August 17-21, 1996.

Introduction

The emerging conventional wisdom espoused in OECD reports, G-7 communiques, and many academic writings is that all advanced industrial nations appear to be experiencing a "jobs" problem: there simply appear to be too few "decent" or "good" employment opportunities to go around. The leading culprits behind these difficulties are typically viewed as the forces of globalization and technological change and the inability of existing institutional structures to adequately adapt to the "new" and rapidly evolving environment. The jobs problem manifests itself somewhat differently in different countries. In the flexible wage United States, private employment growth remains buoyant but there have been dramatic increases in wage inequality, slow overall real wage growth, and substantial declines in real earnings for the less-educated. In generic "inflexible wage/generous welfare state Europe" employment growth has stagnated and unemployment rates have ratcheted up over successive business cycles generating persistent long-term unemployment concentrated among the less-educated and new labor force entrants. The typically unconventional Paul Krugman (1994, p.71) has summarized the hard-headed version of these now conventional views by noting that "the European unemployment problem and U.S. inequality problem are two sides of the same coin" in which markets will tend "to produce increasingly unequal outcomes, or to produce persistent high unemployment if this tendency is repressed."

This interpretation of labor market developments in OECD nations over the past two decades does appear to contain substantial germs of truth. Much research strongly suggests that labor demand has shifted against the less-skilled and the disadvantaged in most advanced economies. Unemployment and nonemployment rates among the less-educated have increased by a sizable amount throughout the OECD. Long-term trends towards declining wage inequality and narrowing wage differentials by skill stopped by the early 1980s. In the United States and United Kingdom wage inequality has dramatically increased since the end of the 1970s, while more modest expansions are apparent in many other countries.

But reality appears to be somewhat messier than a broad explanation focusing solely on a collapse in demand for the unskilled differentially affecting "flexible" and "inflexible" wage economies. There is much heterogeneity in the experiences of nations within the inflexible labor market category. Unemployment among the unskilled appears to be lower in some inflexible wage nations (e.g., Norway, Germany) than in more flexible Britain and Canada. Labor market adjustments to changes in the relative demand for skill also depend on education and training policies, macroeconomic policies and experiences, the effectiveness of active labor market policies, and the nature of institutional wage setting in a manner more complicated than suggested by a simple diabolical trade-off between inequality and unemployment.

In this paper I assess the importance of globalization factors (increased trade and immigration, particularly from newly industrialized and less-developed nations) and skill-biased technological change (the computer revolution) in shifting employment opportunities in advanced nations against less-educated workers and those doing more routinized tasks and towards more-educated workers. I first examine how the decline in the demand for the less-skilled has manifested itself in the United States where changes in market forces can affect labor market outcomes in a relatively unfettered manner. I then briefly contrast the U.S. experience with that of other OECD nations. Shifts in the demand for skills appear similar across countries so that differences in outcomes appear related to differences in the supply of skills and labor market institutions. I conclude with some speculations on how improvements in macroeconomic performance might affect the situation of the disadvantaged and on the effectiveness of alternative policies to spur employment opportunities for the less-skilled.

Rising Inequality in U.S. Labor Market Outcomes

The inequality of economic well-being has increased substantially along many dimensions in the United States over the past two decades. The enormous disparities in the fortunes of American families in recent years have largely been associated with labor market changes that have increased overall wage inequality and shifted wage and employment opportunities in favor of the more-educated and more-skilled. These changes have been carefully documented by many researchers (of disparate ideological stripes) using a wide variety of publically-available data sets.[1] While much debate exists concerning the causes of rising inequality, there is substantial agreement concerning the "facts" that need to be explained.

Recent broad changes in the U.S. labor market outcomes can be summarized as follows:

- **From the 1970s to the mid-1990s wage dispersion increased dramatically for both men and women reaching levels of wage inequality for men that are probably greater than at any time since 1940.** The weekly earnings of a full-time worker in the 90th percentile of the U.S. earnings distribution (someone whose earnings exceeded those of 90 percent of all workers) relative to a worker in the 10th percentile (someone whose earnings exceeded those of just 10 percent of all workers) grew by approximately 36 percent for men and 30 percent for women from 1979 to 1995 (OECD, 1996). Earnings inequality has expanded even more rapidly if one includes consideration of the very top part of the distribution (the upper 1 percent).

- **Wage differentials by education and occupation increased.** The college/high school wage premium doubled for young workers. The labor market returns to years of formal schooling, academic achievement as measured by test scores, workplace training, and computer skills greatly increased in the 1980s and early 1990s.

- **Wage dispersion expanded within demographic and skill groups.** The wages of individuals of the same age, education, and sex, working in the same industry and occupation, are much more unequal today than ten or twenty years ago.

- **The real earnings of less-educated and lower-paid workers appear to have declined relative to those of analogous workers two decades ago.** The decline in the real and relative wages of less-skilled workers has not been offset by increases in non-wage employee benefits or increased chances of holding a job. Nonemployment rates for less-educated males have increased over the past two decades and the official employment and unemployment numbers understate this rise since the burgeoning population of those incarcerated (over 1.5 million in 1995) is not included the civilian noninstitutional population (Freeman, 1996a).

- **Increased cross-sectional earnings inequality has not been offset by increased earnings mobility.** Permanent and transitory components of earnings variation have risen by similar amounts (Gittleman and Joyce, 1996; Gottschalk and Moffitt, 1994). But this implies that year-to-year earnings instability has also increased substantially over the last two decades.

- **These labor market changes have translated into a large widening of the family income distribution as the earnings of husbands and wives have become more positively correlated over time** (Burtless and Karoly, 1995). While pre-tax money income is a noisy measure of economic well-being, increased inequality is also apparent when one directly examines consumption and accounts for in-kind benefits and government transfers (e.g., Cutler and Katz, 1991; U.S. Department of Labor, 1995a).

Many researchers conclude that a driving force behind these changes has been a large increase in the gap between the rate of growth of the relative demand for more-skilled workers and the rate of growth of the supply of such workers. Katz and Murphy (1992) emphasize both the deceleration in the rate of growth of the relative supply of college graduates since the early 1980s (associated with the entry of baby bust cohorts and increased unskilled immigration to the United States) and a possible acceleration in the rate of growth of the relative demand for the more-educated (from skill-biased technological change and increased international integration.) An alternative (or complementary) story for rising inequality has emphasized changes in U.S. labor market institutions such as the decline in unions, the erosion of the real value of the minimum wage, and the weakening of pay-setting norms that have historically served to compress the wage structure (e.g., Freeman, 1996b).

The recent substantial declines in the relative and real position of less-skilled U.S. workers reflects a striking break from a historical pattern of widely shared rising prosperity. For much of the twentieth century natural market forces of rising output per worker and a rapidly declining ratio of less skilled to more skilled workers as well as pressures from wage-setting institutions seemed to promise a bright future for the less educated and access to further education and upward mobility for their children. To understand what changed over the past two decades, one needs to study movements in the supply and demand for skills over a long historical period rather than just looking at recent events in isolation as has been the tendency of much recent research and discussion.

The Supply of and Demand for Skills in the United States, 1940–95

It is not possible to measure the overall supply of "skills" since much of what is valued in the labor market is unobservable in existing data sets. But reasonably consistent data on the employment and earnings of workers by educational attainment for representative national samples are available since 1940. I use college graduates as a proxy group of highly-skilled workers and those without college degrees as a proxy for less-skilled workers.

Table 1 presents information on trends in the relative employment and earnings of U.S. college graduates from 1940 to 1995.[2] The first column of panel A of Table 1 displays the rapid secular growth of the college graduate share of employment which more than quadrupled from 1940 to 1995. The first column of panel B compares the annualized rate of increase in the relative employment of college graduates across time periods. The rate of growth of the relative supply of college workers accelerated substantially in the 1970s with the enrollment of baby boomers and incentives from the Vietnam War to enter and remain in college. It then decelerated in the 1980s and 1990s with the "baby bust" cohorts and "apparent" decline in the return to college education in the 1970s. The expansion of the supply of college graduates appears to have modestly outstripped the growth in demand from 1940 to 1980 with a decline in the college plus/high school wage ratio (adjusted for demographic composition) from 1.64 in 1940 to 1.48 in 1980. But this pattern has sharply reversed itself since 1980 with the wage ratio reaching 1.76 in 1995. Table 1 also illustrates that the college graduate share of the overall wage bill (wages per hour times hours worked) increased from 0.11 in 1940 to 0.39 in 1995.

How much of the difference behavior of the college wage premium in the 1980s–90s from the 1970s and earlier periods reflects an acceleration in the demand for skills (possibly from globalization and computerization) and how much is the consequence of a slowdown in the rate of growth of supply? This is a difficult question to definitively answer which depends on the degree of substitutability in aggregate production of college and non-college workers. If college and non-college workers are not very substitutable in production, then changes in the rate of growth of the relative supply of workers by education require large swings in the relative wages of college workers to induce firms to accommodate changes in the skill mix.

Table 1

Relative Employment and Wages of College Graduates, United States, 1940–1995

A. College Employment and Wage Bill Shares: 1940–1995

	Employment Share	Hours (FTE) Share	Wage Bill Share	College/High School Wage Ratio
1940 Census	.060	.061	.120	1.64
1960 Census	.101	.106	.164	1.49
1970 Census	.134	.138	.215	1.59
1980 Census	.192	.204	.281	1.48
1980 CPS	.199	.209	.285	1.48
1990 CPS	.246	.261	.373	1.72
1995 CPS	.262	.227	.394	1.76

B. Changes in College/Non-College Log Relative FTE, Wage Bill, Wages, Supply: 1940-1995 100 times Annual Log Changes

	Relative Hours (FTE)	Relative Wage Bill	Relative Wage	Relative Supply
1940–1960	2.99	1.81	-0.50	2.31
1960–1970	3.43	3.35	0.67	2.68
1970–1980	4.69	3.56	-0.73	4.29
1980–1990	2.88	3.99	1.52	2.48
1990–1995	2.53	2.28	0.42	1.86

See notes on page 45

While much uncertainty exists concerning the aggregate elasticity of substitution between more and less-educated workers, the existing evidence suggests it may be close to 1 which implies that changes in the relative demand for college graduates are approximately equal to changes in the relative wage bill share of college graduates. Panel B of Table 1 indicates that the rate of growth of the college graduate share of the wage bill accelerated in each decade from the 1940s to the 1980s with some evidence of deceleration in the 1990s (but changes in coding of education questions in the

Current Population Surveys and the short period covered suggest caution in drawing conclusions from the 1990s data). Under this scenario the differences in the behavior of the college/high school wage differential in the 1970s and 1980s (illustrated in the third column of panel B of Table 1) are largely driven by the deceleration in the growth of the relative supply of college graduates in the 1980s (shown in fourth column of panel B of Table 1). Somewhat higher estimates of the degree of substitutability between college and non-college workers from Katz and Murphy (1992) suggest roughly equivalent contributions of demand acceleration and supply deceleration for the 1980s rise in skill premia.

Thus the rate of growth of the demand for more-educated workers appears to have been increased sharply since the 1960s. Autor, Katz, and Krueger (1997) show that this acceleration in the rate of growth of demand for college workers has been almost entirely the result of skill upgrading within detailed industries rather than from changes in the inter-industry mix of employment. Within-industry skill upgrading in the traded goods sectors could arise from increased outsourcing abroad of less-skilled tasks, but similar patterns in non-traded goods and services strongly suggests the role of skill-biased technological change rather than trade pressures being dominant. The acceleration in the growth of the demand for more-educated labor was offset in the 1970s with rapid increases in supply of college graduates, but this was not the case in the 1980s.

Globalization and the U.S. Labor Market

The trade hypotheses for the declining position of less-skilled workers in advanced industrial nations focuses on reductions in trade barriers and improvements in manufacturing productivity in less-developed countries (LDC's). These developments are viewed as having led to a surge in imports of manufactured goods intensive in unskilled labor from LDC's that have directly displaced many unskilled workers from manufacturing jobs in advanced nations and more generally created downward pressure on less-educated workers. Stephen Nickell's (1996) paper for this meeting provides an excellent summary of the observable implications of the trade hypothesis and the existing evidence on its quantitative importance. I'd like to make several additional points.

U.S. manufacturing imports from LDC's as a share of GDP expanded from 0.8 percent in 1970 to 2.3 percent in 1980 to 2.8 percent in 1990 and to 4.1 percent in 1997 (Borjas, Freeman, and Katz, 1997). Sachs and Shatz (1996) simulate the effects of increased imports from LDC's over the 1979 to 1990 period on U.S. relative labor demand assuming these imports potentially displaced U.S. production using factor ratios similar to U.S. domestic production in the same detailed industry. Borjas, Freeman, and Katz (1996) perform a similar factor content calculation for overall changes in U.S. net imports in the 1980s. Both studies conclude that actual trade flows reduced the relative demand for high school workers relative to college workers by a 1 to 2 percent which is quite small relative to the 25 percent decline in the relative supply of U.S. non-college workers over the same period (as indicated in Table 1). Wood (1994) gets much larger estimates of how much the factor content of LDC trade reduces the relative demand for less-educated workers in advanced nations by (1) focusing on the impact of LDC trade on the relative demand for a more limited group of less-skilled workers (high school dropouts) and (2) by assuming that in the absence of imports from LDCs the advanced nations would increase domestic production in import-competing industries using technologies (factor ratios) similar to those found in low-wage countries. Nickell (1996) observes that the approach of using contemporaneous labor input coefficients from advanced nations is a likely to understate the displacement of less-skilled workers and Wood's approach is likely to overstate the magnitude. A more plausible approach than either of these alternatives might be to assume the U.S. would produce such goods with similar factor ratios to those found in the historical U.S. experience of the same industry before imports expanded from low-wage countries. Assuming the U.S. went back to producing goods imported from low-wage countries (e.g., shoes, apparel, etc.) with factor ratios such as prevailed in the United States during the early 1970s and assumed completely inelastic domestic demand for these products, Borjas,

Freeman, and Katz (1997) estimate that the elimination of all imports of manufactured goods from LDCs in 1995 would have reduced the relative demand for high school dropouts by 10 percent and the relative demand for non-college workers by 4 percent. Thus under these extreme assumptions LDC trade could account for one-fifth to one-third of the decline in relative wages of U.S. high school dropouts since 1980 and no more than one-sixth of the decline for the broader group of non-college workers.

The rapid increase in U.S. trade with developing countries of the past fifteen years is likely to continue and possibly even accelerate into the future. An alternative approach to looking backwards to assess the impact of trade on the labor market is taken by Lawrence and Evans (1996) who use a simple computable general equilibrium model to simulate the effects of a five-fold increase in imports of manufactured goods from developing countries. Their simulation involves a balanced trade expansion that completely eliminates U.S. production in industries competing with LDC imports (basic manufacturing industries such as clothing, primary metals, rubber, leather, etc.) and displace half of U.S. manufacturing workers who are re-employed in remaining high-tech manufacturing sectors and non-trade sectors. They find that such a massive change in low-skilled imports would reduce the relative demand for high school workers by approximately 7 percent relative to college workers possibly reducing their relative wages by 5 to 7 percent. This scenario reflects a substantial decline in the relative demand for less-skilled workers, but such a decline over a 10 to 20 year period would be modest relative to trend declines in the relative supply of such workers of 25 to 30 percent a decade.

The basic insight from the Lawrence and Evans simulation is that if the U.S. completely moves out of the production of low-skill traded goods, then the downward pressures of factor price equalization will no longer operate on less-educated American workers. In this case, the floor on wages for such workers are not unskilled wages in China or India, but what they can earn in U.S. exports or domestic industries. But their seemingly extreme scenario makes the strong assumption that the current traded goods/nontraded goods margin remains fixed. An alternative scenario is that many potentially tradeable services (routine operations in financial services, insurance, telemarketing) will become subject to competitive pressure and be potentially transferable to low-wage countries. Thus the future relative demand shock against less-skilled workers could be much larger than indicated by Lawrence and Evans. But their simulation strongly suggests displacements of unskilled workers from trade in manufactures in the past fifteen years (which have certainly not reduced U.S. manufacturing by over 50 percent in the last decade) could not explain more than a few percentage points of the approximately 15 to 20 percentage point decline in the relative wages of the unskilled.

Thus a reasonable conclusion from existing empirical studies seems to be that trade with developing countries has reduced the demand for unskilled labor in developed countries by a noticeable but modest amount. Globalization not only may induce an inward shift in the demand curve for less-skilled labor in advanced nations, but it also makes labor more easily substitutable—through trade and investment flows—across national borders. In a more open economy, employers and final consumers can substitute foreign workers for domestic workers more easily either by investing abroad or importing the products of foreign workers. Thus reduced international trade and investment barriers among developed countries as well as with developing countries tends to make labor demand more elastic as emphasized by Rodrick (1997).

This flattening of the labor demand curve may operate to reduce the historical bargaining power of workers in industries with product market rents in the same way that reductions in transportation costs creating national product markets in the nineteenth century and early twentieth put pressure of workers in previously localized product markets and spurred the creation of national industrial unions as a worker response to the employer "whipsaw" advantage. The creation of worldwide unions of workers by industry appears quite unlikely so that the wages and benefits of less-educated workers in traditionally high-wage sectors are likely to come under pressure as employers can implicitly and explicitly threaten to relocate production across national borders. Increased volatility in labor demand and earnings might also arise as employers and customers can more rapidly take advantage of cost-shifts across nations and previously insulated workers are exposed to shocks in world markets. Borjas and Ramey (1995) provide initial evidence of modest reductions in labor rents for workers in traditional high-wage manufacturing industries increasingly subject to import competition and outsourcing pressure.

Increased immigration is another mechanism through which globalization has affected the U.S. labor market. In 1980 6.4 percent of the American work force was foreign-born; in 1995, 9.8 percent of the work force was foreign-born. Growth in the number of immigrant workers accounted for 20 percent of the growth of the U.S. work force in the 1980s. A disproportionate number of immigrants have less than a high school education, increasing the supply of less educated workers and potentially contributing to the observed decline in their relative pay.

There are two ways to examine the effect of immigration on labor market outcomes. Exploiting the fact that immigration is geographically concentrated in the United States, area analyses contrast the level or change in immigration by area with the level or change in the earnings (or employment) of non-immigrant workers. Area studies have generally found that immigration has had only a slight effect on native outcomes (Borjas, 1994; Borjas, Freeman, and Katz, 1997): native earnings or employment do not differ much between the gateway areas that receive immigrants and other parts of the country. But this approach has some weaknesses that lead to understatements of the possible adverse effects on unskilled immigration on the wages of unskilled U.S. workers. Immigrants are likely to be attracted to geographic areas that are booming making it difficult to uncover the immigration effect without good controls for local demand conditions. Much evidence suggests the migration of less-skilled native workers responds to immigration-induced changes in local labor market outcomes thereby diffusing over a broader geographic area the impact of immigrants on the relative supply of skills. Capital may also relocate within the U.S. to immigrant-receiving areas; it would not be surprising to learn that apparel industry investment in Los Angeles or New York may be affected by a steady stream of low-wage immigrant labor supply.

The factor proportions approach to seeing how immigration affects the job market assumes that the effects of immigration and trade are sufficiently diffused across areas due to native migration or capital responses that it is best to examine the effect of immigration through its effect on the national supplies of labor with different skills. Borjas, Freeman, and Katz (1997) find that post-1979 immigration increased the relative supply of U.S. high school dropouts by approximately 15 percent in 1995 and estimate that the increased pace of unskilled immigration from 1980 to 1995 may have reduced the relative earnings of the American high school dropouts by 3 to 4 percent (of an approximately 11 percent overall decline in relative earnings of this group over this period).

Immigration appears has been fairly important in reducing the pay of high school dropouts, while the direct effects of immigration and trade have contributed modestly to the falling pay of high school graduates versus college workers in the United States. More indirect threat effects of trade and potential trade and outsourcing on the wages of less-skilled Americans could have been larger but are difficult to measure.

Skill-Biased Technological Change and the Wage Structure

The skill-biased technological change explanation for rising wage inequality has particular appeal to many economists. The continued increase in the relative utilization of nonproduction workers and more educated workers within detailed industries and within establishments in the United States despite the rising relative price of these groups during the 1980s and 1990s suggests strong within-industry and within-establishment demand shifts favoring the more-educated. Similar within-industry increases in the proportion of "skilled" workers are apparent during the 1980s in other OECD nations and the increases in skill demand are most rapid in the same set of manufacturing industries across OECD economies (Berman, Machin, and Bound, 1996). Skill-biased technological change is the "natural" name for economists to attach to unexplained within-sector and within-firm growth in the demand for skill.

The diffusion of computers and related technologies represents a prime suspect for a widespread recent technological change that could lead to major changes in the demand for skill within industries and firms. One approach to measuring the spread of computer technology is by the fraction of workers who directly use a computer keyboard. Although this approach misses workers who use devices that contain microprocessors not operated by keyboards, it measures a particularly prevalent form of computer technology. Table 2 reports the percentage of U.S. workers who report using a computer keyboard at work in 1984, 1989, and 1993. The prevalence of computer use at work increased almost linearly from one-quarter of the work force in 1984, to over one-third in 1989, and to nearly one-half in 1993—an average increase of 2.4 percent of the workforce per year. Similar growth in computer utilization with a slight lag is apparent in Britain, Canada, France, and Germany.

Much evidence exists beyond indirect inferences indicating that capital and new technologies are relative complements with "more-skilled" workers.[3] Krueger (1993) and Autor, Katz and Krueger (1997) find a substantial and growing wage differential associated with computer use from 1984 to 1993. Allen (1996) shows that educational wage differentials grew more rapidly in the 1980s in industries that were intensive in high-technology capital and those having more rapid growth in scientists and engineers as a share of the work-force. Berman, Bound, and Griliches (1994) find that skill upgrading in U.S. manufacturing sectors was strongly positively correlated with computer investment and R&D intensity during the 1980s.

But there are reasons to be a bit somewhat skeptical that skill-biased technological change is the sole reason for divergent patterns in wage structure changes in the 1980s–90s than in earlier periods. Strong evidence of capital-skill and technology-skill complementarity is apparent throughout the twentieth century in periods of relatively stable or narrowing educational wage differentials as well as the more recent period of rapidly rising wage differentials. Goldin and Katz (1996a,b) find that the introduction of continu-

Table 2

Percentage of Workers in Various Categories Who Directly Use a Computer at Work

	October 1984	October 1989	October 1993
Use a computer			
All Workers	25.1	37.4	46.6
Gender			
Male	21.6	32.2	41.1
Female	29.6	43.8	53.2
Education			
Less than HS	5.1	7.7	10.4
High School	19.2	28.4	34.6
Some College	30.6	45.0	53.1
College +	42.4	58.8	70.2
Race			
White	25.8	38.5	48.0
Black	18.6	28.1	36.7
Age			
Age 18–24	20.5	29.6	34.3
Age 25–39	29.6	41.4	49.8
Age 40–54	23.9	38.9	50.0
Age 55–64	17.7	27.0	37.3
Occupation			
Blue collar	7.1	11.2	17.1
White collar	39.7	56.6	67.6
Union Member			
Yes	19.9	31.8	39.1
No	25.3	37.7	46.9
Hours			
Part-time	14.8	24.4	29.3
Full-time	29.3	42.3	51.0
Region			
Northeast	25.5	37.6	46.9
Midwest	24.3	36.6	46.7
South	23.2	36.6	45.0
West	28.9	39.7	48.8

Data source: October 1984, 1989, and 1993 Current Population Surveys. Sample sizes are 61,704, 62,748 and 59,852 in 1984, 1989 and 1993 respectively. Results are weighted by CPS sample weights. Sample includes workers ages 18–64 who were working, or with job but not at work in previous week.

ous-process and batch methods of production and the adoption of electric motors greatly increased the relative demand for nonproduction workers and more-educated production workers from 1909 to 1929, but that wage differentials by skill did not increase during the period. The rapid increase in the supply of skills arising from the high school movement may have prevented wage inequality from rising in the face of what appears to be a skill-biased technological revolution.

While existing research has convincingly shown that skill-biased technological has been the main force raising the relative demand for more-skilled workers over the long run, evidence of an acceleration in the pace of skill-biased technological change in the 1980s and 1990s is largely indirect. New work by Autor, Katz, and Krueger (1997) does find that indicators of computer usage, computer capital per worker, and computer investment as a share of total investment are higher in industries with substantial accelerations in skill upgrading in the 1970s and 1980s versus the 1960s than in industries with little or no such acceleration. While this may not be a "causal" relationship, whatever is driving increases in the rate of growth of demand for skilled labor over the past twenty-five years seems to be concentrated in the most computer-intensive sectors of the economy.

In summary, both a deceleration in the rate of growth of the supply of skills and an acceleration in the relative demand for skills contributed to the substantial growth of wage inequality in the United States since the end of the 1970s. The demand acceleration associated with an increased pace of skill-biased technological change probably started earlier than the 1980s but was held in check by the rapid growth of the supply of college graduates with the labor market entry of baby boom cohorts. Trade has not yet been as important as technological change in the reduced relative demand for the less-skilled but could become much more important in the future. The impacts of immigration have been particularly concentrated on the least educated American workers. The weakening of unions, the minimum wage, and pay-setting norms allowed these market changes in supply and demand to greatly affect the wage structure and have probably been important also in the large increases in within group inequality and earnings instability.

Wage Inequality and Unemployment in Other Advanced Nations

The unemployment rates of unskilled workers have increased in essentially all advanced nations over the last two decades suggesting a common shift in labor demand against the less-skilled (Nickell, 1996). But wage structure changes have differed substantially among OECD nations with countries with more decentralized wage determination and less government constraints on wage setting showing much larger increases in wage inequality. Table 3 measures changes in inequality in terms of the log of the ratio of the earnings of the top decile to the bottom decile of earnings for males from 1979 (or the earliest year available) to 1994 (or the latest year available). The data show that the United States and United Kingdom had by far the biggest increases in inequality and are the only nations where wage inequality has persistently been increasing over this period. The pattern of declining wage inequality apparent throughout the OECD (except for the United States) in the 1970s ceased in the 1980s and 1990s in almost all nations (with Germany and Norway as exceptions). Substantial increases in wage dispersion are apparent in New Zealand with substantial product market and labor market deregulation over the last decade and in Italy after 1980 with the abolition of an automatic cost-of-living index favoring low-wage workers and the ending of synchronization in bargaining across industries.

Can international differences in changes in relative wages in the 1980s be explained by differences in supply, demand, and institutions? Why did inequality increase more in the United States and the United Kingdom than in most other advanced countries?

Labor demand factors do not explain much of the differential growth of wage inequality or educational earnings differentials among countries in the 1980s. All advanced countries experienced large, steady shifts in the industrial and occupational structure of employment towards sectors and job categories that use a greater proportion of more-educated workers.

Differential growth in the supply of workers by level of educa-

Table 3

Trends in Wage Inequality for Males, Selected OECD Countries, 1979 to 1994[a]

Log of ratio of wage of 90th percentile earner to 10th percentile earner

Country	1979	1984	1989	1994	Change from earliest to latest year
Australia	1.01	1.01	1.03	1.08	0.07
Austria[b]	0.97		1.00		0.03
Canada[c]	1.24	1.39	1.38	1.33	0.09
Finland[d]	0.89	0.92	0.96	0.93	0.04
France	1.22	1.20	1.25	1.23	0.01
Germany[e]		0.87	0.83	0.81	-0.06
Italy	0.83	0.83	0.77	0.97	0.14
Japan	0.95	1.02	1.05	1.02	0.07
Netherlands[f]		0.92	0.96	0.95	0.03
New Zealand[g]		1.00	1.12	1.15	0.15
Norway[h]	0.72	0.72	0.77	0.68	-0.04
Sweden[i]	0.75	0.71	0.77	0.79	0.04
United Kingdom	0.90	1.02	1.12	1.17	0.27
United States	1.16	1.30	1.38	1.45	0.29

[a] The samples generally consist of full-time workers, with the exceptions of Austria, Italy, and Japan. See OECD (1996, pp. 100-103) for details on the samples and earnings measures.

[b] Data for Austria in the 1979 column are for 1980.

[c] Data for Canada are for 1980, 1986, 1990, and 1994

[d] Data for Finland are for 1980, 1983, 1989, and 1994.

[e] Data for Germany are for 1983, 1989, and 1993.

[f] Data for the Netherlands are for 1985, 1989, and 1994.

[g] Data for New Zealand are for 1984, 1990, and 1994.

[h] Data for Norway are for 1980, 1983, 1987, and 1991.

[i] Data for Sweden are for 1980, 1984, 1989, and 1993.

Source: OECD (1996), Table 3.1, pp. 61-62.

tion, in contrast, contributed to the greater rise in education wage differentials in the United States than in other countries in the 1980s. In the 1970s the supply of highly educated workers increased rapidly in all OECD countries. The increase was more rapid than the shifts in demand favoring educated workers, so that skill differentials narrowed in every country (Freeman and Katz, 1994). In the 1980s, however, while the educational qualifications of workers continued to rise in all countries, the growth of the college-educated work force decelerated sharply only in the United States. Continued rapid expansion of the college-educated work force in Canada, for instance, helps explain the more modest rise in the relative earnings of college graduates in Canada than in the United States (Freeman and Needels, 1993). Countries whose education differentials did not increase much during the 1980s (e.g., the Netherlands and Germany) maintained their 1970s rate of growth in the supply of more educated workers into the 1980s.

Given comparable changes in demand across countries, differences in the growth of relative supply help account for country differences in the growth of skill differentials. Still, they cannot explain the bulk of country differences in levels or changes in inequality. Differences in labor market institutions among countries and changes in those institutions in the 1980s and 1990s also influenced the pattern of wage inequality. But while countries with less "flexible" wage setting institutions had little or no growth in wage inequality. This "inflexibility" did not necessarily always translate into higher unemployment for the unskilled (Freeman and Katz, 1994; Nickell, 1996). Countries that combine institutional wage-setting with education and training systems that invest heavily in non-college workers such as in Germany and Norway did not do as poorly in terms of unemployment as one might have expected from their degree of "rigidities." Firms in these nations seem to treat college-educated and non-college workers as much closer substitutes in production than do U.S. or British firms. Technology and trade shocks do not generate as much pressure for wage structure changes in these countries, since workers are not sharply differentiated by skill. Buffering the earnings of the less educated by institutional wage-setting works best when accompanied by institutions that augment their skills as well.

Nevertheless, the experiences of other European nations (e.g., France and maybe Sweden in the 1990s) suggest that inflexibilities in wage setting combined with generous welfare state systems might make it very difficult to recover from macroeconomic shocks that generate sharp increases in unemployment. A long period of high actual unemployment leads to an increase in the proportion of long-term unemployed. If the long-term unemployed either lose skills, or become less effective in search, their effect on bargaining will decrease, leading to higher wages given unemployment, and thus a higher natural rate of unemployment (Blanchard, 1991). Higher employment protection in Europe may also decrease the weight of unemployment on the wage set in bargaining between employed workers and firms.

There may be sociological factors at work as well, which in effect increase the reservation wage of some of the unemployed. A long period of unemployment alters society's attitudes towards the unemployed. It becomes more socially acceptable to be unemployed and to use existing benefits to their utmost (Lindbeck, 1995). Families adjust to unemployment by increasing family insurance possibly letting youth stay at home longer. Government policies adapt with pressure for more generous programs aimed at helping the unemployed. While these changes decrease the pain of unemployment, they are also likely to increase the natural rate of unemployment. Thus the European unemployment problem may have important hysteresis components beyond a simple unicausal explanation focusing on the collapse of demand for the less-skilled. But the growth of the welfare state in this situation may become less sustainable as the tax base becomes somewhat more footloose with increased globalization implying greater international capital mobility and high-skilled labor mobility.

Macroeconomic Performance and the Disadvantaged

Economic growth and strong macroeconomic performance has traditionally been seen as the dominant source of gains for the poor and of improved economic opportunities for individuals located throughout the income distribution. During the 1960s, for example, rapid economic growth and a relatively stable economy dramatically reduced the share of Americans living in poverty, by 10.3 percentage points, and moved many families into a thriving middle class. During the 1970s, a period of unstable economic conditions and slow growth, poverty rates were relatively constant. Although the recession of the early 1980s saw an almost 4 percentage point increase in poverty, an analyst in 1983 using the historical relationship (from 1959 to 1983) between macroeconomic performance (as measured by the ratio of the poverty line to median or mean family income, the inflation rate, and the prime age male unemployment rate) and poverty would have expected the sustained economic expansion of 1983 to 1989 to lead to another surge in the well-being of the nation's disadvantaged.

But, as Figure 1 illustrates, the anticipated benefits for the poor of the 1980s expansion failed to fully materialize and a similar pattern appears through the early part of the current expansion. If the "trickle down" mechanism had been as effective in the 1980s and 1990s as in the past, the poverty rate would have been 4.0 percentage points lower than actually observed in 1995 (9.8 percent rather than 13.8 percent). Figure 2 shows an even larger "unanticipated" increase in the child poverty rate of 7 percentage points in 1995 (21.3 percent as opposed to a predicted 13.8 percent rate). The story is equally true of overall family income inequality—again the 1980s and 1990s are a sharp break from historical experience. Similar results occur if one looks at family consumption rather than income and if one accounts for in-kind benefits such as Medicaid (Cutler and Katz, 1991).

While stronger macroeconomic performance is still a necessary condition for improving the economic opportunities for the disadvantaged, it is unlikely to be sufficient barring a supply-side miracle that more than doubles the trend productivity growth rates in advanced nations. The trickle down mechanism is less effective than in the past with a smaller share of increased growth translating into increased demand for the services of the disadvantaged and non-college workers. Thus secular structural labor market shifts

Figure 1

U.S. Actual and Predicted Poverty Rates, 1959-95

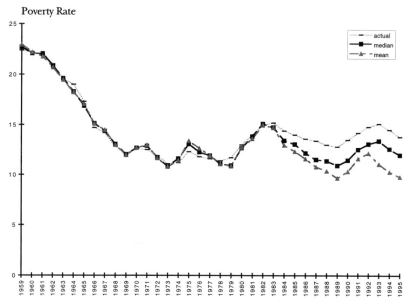

Figure 2

U.S. Actual and Predicted Child Poverty Rates, 1959-95

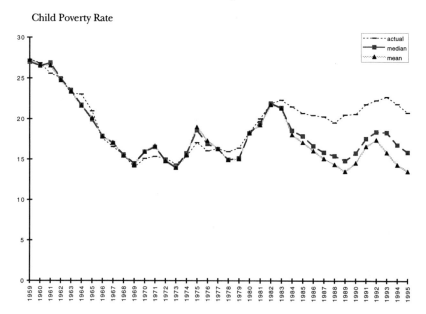

against the less-skilled, changes in labor market institutions, and the cumulative process of changes in families and neighborhoods have served to offset the traditional benefits of economic growth for disadvantaged Americans and similar processes may be beginning in some high unemployment regions of Europe. Rapid economic growth and tight labor markets need to be complemented with work-force preparation strategies that better enable those without college degrees and from poor backgrounds to take advantage of emerging opportunities and possibly with policies to subsidize the employment of the less-skilled.

Market incentives for increased individual educational investments and skill upgrading can play some role in alleviating growing inequality in the United States. The large increase in the college wage premium in the 1980s has been associated with an increase in college enrollment rates from 49 percent of high school graduates in 1980 to more than 60 percent in the early 1990s. But the process of supply adjustment can take many years, and many disadvantaged individuals face financial and informational barriers to pursuing further education and training.

Much recent work has evaluated the effectiveness of alternative active labor market policies for improving the labor market prospects of the disadvantaged. Policies to increase years of schooling for those from low-income families appear to have a high returns (Card, 1994). Public sector-sponsored training programs have a mixed record with strong positive returns for disadvantaged adults (particularly adult women) and more disappointing results in evaluations of programs for disadvantaged out-of-school youths (LaLonde, 1995; and U.S. Department of Labor, 1995b). Employer-side wage subsidies (or employment tax credits) that are highly targeted on very specific socioeconomic groups appear somewhat effective for disadvantaged youth but have substantial administrative burdens and may even stigmatize some targeted groups (e.g., welfare recipients and ex-convicts). Policies using an intermediary (a public employment agency, nonprofit training organization, etc.) that combine job development, job search assistance, training, and employment subsidies appear more successful for targeting on specific disadvantaged groups (Katz, 1996). The earned income tax credit (EITC) is also an integral part of a strategy to improve the earnings of less-skilled workers and encouraging movements from dependency to work.

Notes to Table 1

This table is adapted from Tables 1 and 2 and Appendix Table A1 of Autor, Katz, and Krueger (1997). College employment, hours/FTEs (full-time equivalents), and wage bill shares are calculated for samples including all individuals in paid employment (both wage and salary and self-employed workers) during the survey reference week for each Census and CPS sample. 1940-60, 1960-70, 1970-80 changes use data from the 1940, 1960, 1970, and 1980 Census PUMS samples. 1980-90 and 1990-94 changes use data from the 1980, 1990, and 1995 CPS Merged Outgoing Rotation Groups. Wages are imputed for self-employed workers in both the CPS and Census samples using the average hourly wages of wage and salary workers in the same industry-education-year cell. Hours per week for self-employed workers in the CPS samples are imputed using average usual weekly hours for wage and salary workers in the same industry-education-year cell.

The college/high school wage ratio is the (adjusted) ratio in earnings of those with 16 or more years of schooling to those with exactly 12 years of schooling. More specifically, the college/high school wage ratios in Panel A are the exponentiated fixed-weighted average coefficients on dummy variables for those with exactly 16 years of schooling and those with 17 or more years of schooling from cross-section log hourly earnings regressions for wage and salary workers in each years with dummy variables for individual years of schooling (12 years of schooling is the base group), a quartic in experience, 3 region dummies, a part-time dummy, female and nonwhite dummies, and interaction terms between the female dummy and the quartic in experience and the nonwhite dummy. The weights are the shares of college workers with exactly 16 and more than 16 years of schooling in the base year of 1980. The change in log relative supply is the difference between the change in the log relative wage bill and the change in the log relative wage.

End Notes

[1] More detailed references and surveys of the literature can be found in Levy and Murnane (1992), Freeman and Katz (1994) and Freeman (1997).

[2] Table 1 uses data from the 1940, 1960, 1970, and 1980 Census of Population Public Use Micro Samples (PUMS) and from the 1980, 1990, and 1995 Current Population Survey (CPS) Merged Outgoing Rotation Group files. See Autor, Katz, and Krueger (1997) for a more in-depth analysis of changes in the relative supply and demand for skills in the U.S. labor market over the 1940 to 1995 period.

[3] See Hamermesh (1993) for a survey of evidence on capital-skill and technology-skill complementarity.

References

Allen, Steven G. "Technology and the Wage Structure." NBER Working Paper No. 5534, April 1996.

Autor, David, Lawrence F. Katz, and Alan Krueger. "Computing Inequality: Have Computers Changed the Labor Market?" NBER Working Paper No. 5956, March 1997.

Berman, Eli, John Bound, and Zvi Griliches. "Changes in the Demand for Skilled Labor within U.S. Manufacturing Industries: Evidence from the Annual Survey of Manufacturing." *Quarterly Journal of Economics* 109 (May 1994): 367-97.

Berman, Eli, Stephen Machin, and John Bound. "Implications of Skill-Biased Technological Change: International Evidence." Unpublished paper, Boston University, July 1996.

Blanchard, Olivier J. "Wage Bargaining and Unemployment Persistence." *Journal of Money, Credit, and Banking* (1991): 277-92.

Borjas, George J. "The Economics of Immigration." *Journal of Economic Literature* 32 (December 1994): 1667-1717.

Borjas, George J. and Valerie A. Ramey. "Foreign Competition, Market Power, and Wage Inequality." *Quarterly Journal of Economics* 110 (November 1995): 1075-1110.

Borjas, George J., Richard B. Freeman, and Lawrence F. Katz. "Searching for the Effect of Immigration on the Labor Market." *American Economic Review* 86 (May 1996): 246-51.

Borjas, George J., Richard B. Freeman, and Lawrence F. Katz. "How Much Do Immigration and Trade Affect Labor Market Outcomes?" *Brookings Papers on Economic Activity*, 1997, forthcoming.

Bound, John and George Johnson. "Changes in the Structure of Wages in the 1980s: An Evaluation of Alternative Explanations." *American Economic Review* 82 (June 1992): 371-92.

Burtless, Gary and Lynn Karoly. "Demographic Change, Rising Earnings Inequality, and the Distribution of Personal Well-Being, 1959-89." *Demography* 32 (August 1995): 379-405.

Card, David. "Earnings, Schooling, and Ability Revisited." Princeton University, Industrial Relations Section Working Paper No. 311, May 1994.

Cutler, David M. and Lawrence F. Katz. "Macroeconomic Performance and the Disadvantaged." *Brookings Papers on Economic Activity*, 1991:2, 1-61.

Freeman, Richard B. "Why Do So Many Young American Males Commit Crimes and What Might We Do About It?" *Journal of Economic Perspectives* 10 (Winter 1996a): 25-42.

Freeman, Richard B. "Labor Market Institutions and Earnings Inequality." *New England Economic Review* (May/June 1996b): 157-68.

Freeman, Richard B. *When Earnings Diverge*. Washington D.C.: National Planning Association, 1997.

Freeman, Richard B. and Lawrence F. Katz. "Rising Wage Inequality: The United States vs. Other Advanced Countries." In R. Freeman, ed., *Working Under Different Rules*. New York: Russell Sage Foundation and NBER, 1994, 29-62.

Freeman, Richard B., and Karen Needels. "Skill Differentials in Canada in an Era of Rising Labor Market Inequality." In D. Card and R. Freeman, eds., *Small Differences That Matter*. Chicago: University of Chicago Press for NBER, 1993.

Gittleman, Maury and Mary Joyce. "Earnings Mobility and Long-Run Inequality: An Analysis Using Matched CPS Data." *Industrial Relations* (April 1996): 180-96.

Goldin, Claudia and Lawrence F. Katz. "Technology, Skill and the Wage Structure: Insights from the Past." *American Economic Review* 86 (May 1996a): 252-257.

Goldin, Claudia and Lawrence F. Katz. "The Origins of Technology-Skill Complementarity." NBER Working Paper No. 5657, July 1996b.

Gottschalk, Peter and Robert Moffitt. "The Growth of Earnings Instability in the U.S. Labor Market." *Brookings Papers on Economic Activity* 1994:2, 217-72.

Hamermesh, Daniel S. *Labor Demand*. Princeton: Princeton University Press, 1993.

Katz, Lawrence F. "Wage Subsidies for the Disadvantaged." NBER Working Paper No. 5679, July 1996.

Katz, Lawrence F. and Kevin M. Murphy. "Changes in Relative Wages, 1963-

1987: Supply and Demand Factors." *Quarterly Journal of Economics* 107 (February 1992): 35-78.

Krueger, Alan B. "How Computers Have Changed the Wage Structure? Evidence from Micro Data." *Quarterly Journal of Economics* (February 1993): 33-60.

Krugman, Paul. "Past and Prospective Causes of High Unemployment." In Federal Reserve Bank of Kansas City, *Reducing Unemployment: Current Issues and Policy Options*. Kansas City, MO: Federal Reserve Bank of Kansas City, 1994, 49-80.

LaLonde, Robert J. "The Promise of Public-Sector Sponsored Training Programs." *Journal of Economic Perspectives* 9 (Spring 1995): 149-68.

Lawrence, Robert Z. and Carolyn L. Evans. "Trade and Wages: Insights from the Crystal Ball." NBER Working Paper No. 5633, June 1996.

Levy, Frank and Richard Murnane. "U.S. Earnings Levels and Earnings Inequality: A Review of Recent Trends and Proposed Explanations." *Journal of Economic Literature*, Sept. 1992.

Lindbeck, Assar. "Hazardous Welfare-State Dynamics." *American Economic Review* 85 (May 1996): 9-15.

Nickell, Stephen. "Unemployment and Wages in Europe and North America." Unpublished paper, University of Oxford, June 1996.

OECD. *Employment Outlook*. Paris: OECD, 1996.

Rodrick, Dani. *Has International Economic Integration Gone Too Far?* Washington D.C.: Institute for International Economics, 1997.

Sachs, Jeffrey D. and Howard J. Shatz. "International Trade and Wage Inequality in the United States: Some New Results." Unpublished paper, Harvard University, February 1996.

U.S. Department of Labor. *Report of the American Workforce*. Washington: U.S. GPO, 1995a.

U.S. Department of Labor. Office of the Chief Economist. *What's Working (and what's not): A Summary of Research on the Economic Impacts of Employment and Training Programs,* January 1995b.

Wood, Adrian. *North-South Trade, Employment, and Inequality*. Oxford: Clarendon Press, 1994.

Unemployment and Wages in Europe and North America

Stephen Nickell

Institute of Economics and Statistics
University of Oxford
June 1996

This paper was prepared for the 1996 Conference of the Aspen Institute Program on the World Economy, August 17th-21st, 1996, Aspen, Colorado. I am grateful to the Leverhulme Trust (Programme on the Labour Market Consequences of Technical and Structural Change), the Economic and Social Research Council, Brian Bell, Tracy Jones and Daphne Nicolitsas for their help in the preparation of this paper.

Introduction

In the developed world, there have been dramatic changes in the position of unskilled workers over the last two decades. And these are, by and large, changes for the worse. In nearly all OECD countries, unemployment rates among the unskilled have doubled or even trebled since the early or mid 1970s. Furthermore in some OECD countries, the pay of the unskilled has fallen dramatically relative to that of the skilled. But in other countries this has not happened. This has led some to argue that in countries where relative wages have not adjusted, the rise in unskilled unemployment has been particularly severe (see, for example, Krugman, 1994; Freeman, 1995). Indeed, the broad brush picture which is typically presented is one where Europe, with its sclerotic labour markets and rigid relative wages, has suffered from dramatic rises in unskilled unemployment whereas in North America, wage flexibility has led to a dramatic widening of the earnings distribution but a much less severe unemployment problem.

However, perhaps the brush that paints this picture is rather too broad. Sclerotic Norway has less unskilled unemployment than flexible America. Flexible Britain, on the other hand, has had higher unskilled unemployment rates than most other European countries. Low skill workers in inflexible (West) Germany are paid more than twice as much as equivalent workers in the United States and yet their unemployment rates are much the same.

The remainder of this paper is concerned to investigate these issues in some depth. First we look at why the relative position of the unskilled has become so much worse. Then we consider the overall consequences of the relative demand shift and why these differ from country to country. We finish with some general conclusions.

Why Has the Demand for Unskilled Workers Fallen?

During the last two decades, it became increasingly evident that the demand for low skill workers in the OECD countries was falling at a rapid rate. Over the same period, there was a substantial increase in OECD imports of manufactured goods from the developing world. Thus, in the early 1970s, such imports represented just over .5 percent of OECD GDP and by 1992, they had risen to about 2.4 percent.[1] The temporal coincidence of these two events instigated a sequence of investigations into the possibility of a causal relationship. Specifically the aim has been, and still is, to find out whether imports of low-skill intensive manufactures from outside the OECD are displacing significant numbers of low-skill jobs inside it. This is not the only possibility. An obvious alternative is that technological change is systematically eliminating low-skill jobs and that this process, which has been in operation at least since the time of the Luddites,[2] has been moving faster in recent years.

These two hypotheses, that the fall in demand for the unskilled is due either to trade (globalization) or technology, remains the subject of intense debate. Investigating the issue further, we start by setting out the hypotheses in detail.

The Trade Argument

This starts with a rise in productivity in unskilled labour intensive manufacturing in less developed countries (LDCs) combined with a reduction in trade barriers and a continuing increase in the supply of unskilled labour, as workers leave agriculture in LDCs in very large numbers. This group of factors leads directly to a fall in the world price of traded goods which are intensive in unskilled labour (unskilled intensive goods) relative to the world price of skilled intensive products.

In its turn, this will lead to a fall in the demand for unskilled labour relative to skilled labour in developed countries (DCs) and hence to a fall in unskilled wages relative to skilled wages. Of course, in the case where the relative wage of unskilled labour is exogenously fixed, this will tend to generate unskilled unemployment instead. Finally, if the relative wage of unskilled labour does fall, then this will generate a *ceteris paribus* rise in the relative employment of unskilled labour in the non-traded sector.

The Technology Argument

This begins with the notion that OECD technological progress in all sectors is biased against unskilled workers and in favour of skilled workers. This will lead to an increase in the relative demand for skilled workers across the OECD and a decline in the relative wages of unskilled workers (so long as relative wages are flexible). Note that, in this case, there will be a demand shift against the unskilled even in the non-traded sector. This contrasts with the consequences of globalization, where the decline in the relative wages of the unskilled leads to an increase in their relative employment in the non-traded sector.

Further Points on Trade and Technology

First, one rather subtle point. It has been argued (see Leamer 1995) that technological progress biased against the unskilled may have no impact on their relative wage. This idea is based on the fact that the relative wage of the unskilled in any one country depends solely on the world prices of unskilled and skilled intensive products, so long as the country concerned produces and trades in both. Technological progress biased against the unskilled in that country will have no effect on relative wages so long as it has no impact on world prices. But this misses the point of the technology argument. The thrust of this argument is that technical progress is biased against the unskilled across the OECD. World prices are hardly going to remain unchanged in this process, so the above objection does not apply.

A more appealing argument, discussed extensively in Wood (1995), notes that technological progress is not exogenous. One of the driving forces behind innovation may be increased LDC competition. Since this is in the unskilled sector, this may provide a strong incentive for firms in this sector to innovate and thereby economise on unskilled labour. So technical progress biased against unskilled labour may not be exogenous at all but merely another consequence of globalization.

Finally, it is worth noting that the trade argument need not rely on changes in prices. For example, developed countries may shift physical capital to LDCs to make unskilled intensive products which are then exported back to the developed world. This will reduce the relative demand for unskilled labour and hence relative unskilled wages across the OECD without any necessary change in the relative price of unskilled intensive goods (see Sachs and Shatz, 1996).

Having set out the hypotheses, we next turn to the evidence.

The Evidence on the Trade and Technology Arguments

(a) **Prices.** In order for the trade argument to carry any weight, we must observe a rise in the world price of skilled intensive products relative to that of unskilled intensive products. Broad brush evidence is provided by Minford (1996) which we reproduce in Figure 1. Here, he presents the price of LDC manufacturing exports (unskilled intensive) relative to that of DC exports of machinery, transport equipment and services (skilled inten-

Figure 1

Ratio of Developing Countries' Manufacturing Export Prices ($) To Developed Countries' Export Prices of Machinery and Transport Equipment and of Services ($)

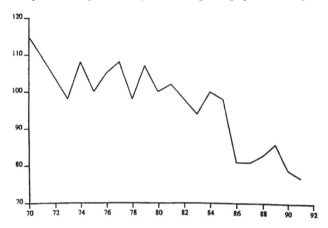

Source: UN International Trade Statistics Yearbook, Various Issues. For services the source is UK Pink Book, export of services price index converted into dollars. The weights are based on shares of non-oil exports and exports of services in UK trade, respectively 0.73 and 0.27. Reproduced from Minford (1996).

sive). The decline from the late 1970s is very marked, being of the order of 30 percent.

More detailed investigations produce less compelling results. Lawrence and Slaughter (1993) consider prices in a large number of manufacturing sectors. They then take weighted averages of these prices first using the percentage of non-production (skilled) workers as weights and second using the percentage of production (unskilled) workers as weights. They find that the former weighted average has risen *less-rapidly* than the latter weighted average in both import and export sectors in the United States, Japan and Germany. Apparently traded goods prices have risen *more slowly* in the skill intensive sectors. However Sachs and Shatz (1994), using much the same data, generate different outcomes, essentially by tidying up some of the series and excluding computers. The problem with these is that their enormous price falls tend to distort the overall picture. Sachs and Shatz end with the result that import prices in the skill intensive sectors have risen more rapidly than in the unskilled sector, as the trade argument implies.

More recently, Krueger (1995) has analyzed US data with a more refined industrial breakdown (4 digit). His basic result is that price growth is negatively related to the proportion of unskilled workers, which is again consistent with the trade argument.

While these results are generally consistent with the trade argument, they are also consistent with causality running in the opposite direction. Thus, we might have a world of imperfect competition in most of these industries, with prices being set by firms at least partly as a mark-up on unit costs. If relative skilled wages rise because of skill biased technical change, then this increase will be passed on to prices, generating a rise in the relative prices of the more skill intensive products.

Overall, therefore, the price evidence is consistent with the trade argument but it is not inconsistent with the technology argument.

(b) **Factor content.** The idea here is to look directly at how much skilled and unskilled labour goes into exports and imports and then to compute the impact of trade on the relative demand for labour in the two categories. If there is no trade, the domestic economy stops producing exports and produces imports instead. This raises the demand for unskilled labour because imports are more unskilled intensive.

The first table in Wood (1994) provides a clear example of how this works and is reproduced below. Sachs and Shatz (1994) use US factor proportions in 51 manufacturing sectors. To get from trade to no trade, they add in the skilled and unskilled labour arising from those sectors where the US is a net importer and subtract out the skilled and unskilled labour in those sectors where the US is a net exporter. The negative effects in the table indicate both that imports are more unskilled intensive than exports *and* that imports are more labour intensive than exports.

The trade effects generated by Wood (1994) are vastly

Table 1

Factor Content Estimates of the Impact of Trade with Developing Countries on the Demand for Labour in Manufacturing in 1990
(percentage difference from counterfactual situation with no trade)

	Sachs and Shatz (1994) (US only)	Wood (1994) (All DCs)
All workers	-5.7	-10.8
Skilled workers	-4.3	0.3
Unskilled workers	-6.2	-21.5
Unskilled minus skilled	-1.9	-21.8

Note: As well as the differences in method of calculation and geographical coverage of the estimates, there are differences in (1) the definition of skilled workers (Sachs and Shatz = nonproduction workers, Wood = workers with postbasic education or training); (2) the definition of manufactured exports (Sachs and Shatz include, Wood excludes, goods with a high natural resource content); and (3) the counterfactual (Sachs and Shatz assume no change from the 1978 net export/output ratio in each sector. Wood assumes no manufactured imports from developing countries and a corresponding reduction in exports).

Sources: Sachs and Shatz (1994, Table 13), Wood (1994, Table 4.9 and p. 151, note 45, but recalculated as a percentage of without-trade, rather than actual, employment).

---

greater than those of Sachs and Shatz (1994), particularly on the demand for unskilled workers. The reason for this is that Sachs and Shatz use US skill proportions when computing the change in the US demand for unskilled labour were she to produce the imports herself. Wood argues that this understates the number of unskilled workers required and that it were better to use the corresponding LDC skill proportions instead. Consider a specific, although *invented*, example. Suppose the US is a net importer of shoes. Suppose further that the LDCs produce standard "low-tech" shoes using a lot of unskilled labour but the US produces expensive, handmade, fashion footwear using mainly skilled workers. The Sachs and Shatz method argues that in the no-trade situation, the US would produce its own standard "low-tech" shoes using the same skill balance as it currently does to produce handmade fashion footwear. The Wood method argues that the US would produce its own standard "low-tech" shoes using the same skill balance as the LDCs currently do to produce the shoes. The latter method clearly generates a much higher demand for unskilled labour in the no-trade situation, exactly as we see in the table.

Which method is right? It is arguable that the standard Sachs/Shatz computation represents an understatement. But it also seems likely that Wood over-corrects. The US would almost certainly be able to produce the "low-tech" standard shoes with less unskilled labour than existing LDCs because they would have better machines and better overall production methods.

So while standard factor content analysis,[3] which generates small effects on unskilled labour demand from trade, probably understates the impact, the Wood corrected effects seem likely to represent an overstatement. Furthermore, even as they stand, the Wood effects are not large. Given that the manufacturing sector in the US, say, employs less than 20 percent of the labour force, it is clear that we can only obtain substantial effects if these manufacturing effects spill over into the remainder of the economy. Wood (1995) argues that they do just that. Indeed making a number of adjustments for trade induced innovations, service sector exports and the production of intermediate inputs by the non-traded sector for the traded sector, he concludes that trade could have lowered the economy-wide relative demand for unskilled labour by 20 percent.

However, this is still not a very big demand shift given that in order to have a significant wage effect, the total demand reduction for the unskilled must outweigh the total supply reduction. In the US, for example, the number of those in the labour force with less than 4 years high school (unskilled) fell from 37.5 percent of the labour force in 1970 to 14.5 percent in 1991. This is a fall of 23 percentage points of the labour force or 61 percent of the original unskilled group. So even the Wood adjusted numbers are nothing like big enough for trade factors alone to generate a fall in demand for the unskilled which can outweigh the fall in supply, and thus generate a significant adverse wage effect.

(c) **Evidence from general equilibrium models of trade.** In order to generate really persuasive estimates of the impact of trade on OECD labour markets, we require a properly calibrated general equilibrium model. Such models are presented in Krugman (1994) and Minford (1996). Krugman sets out a model in which two goods are traded (skill and unskilled intensive) and he is careful to ensure that his model is consistent with the low level of trade between LDCs and the OECD. He then concludes that the existence of this trade explains about 10 percent of the shift of relative unskilled wages in the US or, if he imposes relative wage rigidity, the outcome is about 20 percent of the rise in European unemployment rates.

The model given in Minford (1996) is altogether more complex. There are two countries (North and South) and four goods (agriculture, manufactures, services, domestic) where the first three are traded. "Services" consists of traded services and high-tech manufactures whereas "manufactures" refers to low-tech products. Then there are five factors (unskilled labour, skilled labour, land, capital and materials), the first three of which are immobile. "Levels" of technology differ between North and South so that while the traded goods prices suffice to determine the relative prices of the three immobile factors,[4] there is not full factor price equalization.

The model is calibrated and used to investigate the consequences of a 2.3 percent pa growth in manufacturing productivity in the South. The result is that unskilled wages in the North fall by over 3 percent pa relative to skilled wages. This is somewhat faster than the rate achieved in the United States over the 1980s, so it is a massively bigger trade effect than that generated by Krugman. However, Minford does not report the actual volume of trade between North and South generated by his model. It seems clear that, given its impact, it must be vastly greater than the 2.5 percent of North GDP observed in practice. This is crucial for, as Krugman (1995) points out,

"At the very least, this approach lays down a challenge to economists who claim that trade has had very large effects on wages: can they produce a general equilibrium model of the OECD, with plausible factor shares and elasticities of substitution, that is consistent both with their assertions and with the limited actual volume of trade? If they cannot, they have not made their case" (p. 26-27).

(d) **Composition effects.** In our discussion of the trade and technology arguments, we noted that if trade is the driving force, the rise in the relative pay of the skilled would reduce their employment in the non-traded sector. On the other hand, if technology changes are biased in favour of the skilled in all sectors, then the relative demand for their services would rise in all sectors included those insulated from trade.

Berman et al. (1994) investigate this issue for the United States as does Machin (1996) for the UK. Both consider a large number of manufacturing sectors and find that the vast majority of the rise in the proportion of skilled workers is within sector. That is, irrespective of the extent of trade within each sector, the proportion of unskilled workers tends to fall. But this evidence is not that persuasive. Manufacturing sectors are large and manufactures are often traded. If most sectors have a large traded element, a rise in the skilled proportion within the traded part will suffice to ensure that most of the rise in skill will appear as within sector. This interpretation is consistent with a result of Feenstra and Hanson (1995) who find that the rise in import penetration in each sector has a strong positive impact on skill intensity.

However Machin (1996) produces some further evidence which is more compelling. He takes a sample of some 400 establishments which are distributed across all sectors of the economy, including those which are well away from the influence of trade. He then finds the same result. The vast majority of the rise in skill intensity is within establishment. That is, nearly all establishments exhibit increases in skill intensity irrespective of sector. This is inconsistent with a pure trade explanation.

(e) **Technology and wages.** A key problem with the technology explanation for the fall in demand for unskilled workers is that technological change is not exogenous. And if the major driving force behind technological advance is globalization, then we are back to the trade argument. However, while globalization may be a factor, there are other important forces at work. The rapid spread of computers and information technology throughout the non-traded sector is an obvious example. Furthermore, there is clearly an interplay between the accumulation of human capital and technical advance. Increases in human capital will, of itself, raise the return on R and D simply because R and D capital is more productive in the presence of more skilled workers. More innovation will, in its turn, raise the return to education leading to further human capital accumulation. Competition may drive this process more rapidly, but LDC trade provides only a small part of the competition facing companies in the OECD.

The fact that there is a degree of complementarity between skilled labour and technological advance is, of course, hardly sufficient to explain the rise in the relative wages of the skilled in the US or the UK. For a start, it does not reveal why this rise occurred in the 1980s but not in the 1970s or 1960s when technological advance was equally rapid. This must, presumably, have something to do with the rate of increase in the supply of skills. However, there is some direct evidence that technological advance is associated directly with higher wages in the 1980s independent of sector.

The results reported in Krueger (1993) indicate that, in the 1980s, workers in the US who used computers in the workplace obtained a higher wage than those who did not, controlling for qualifications, experience and numerous other observed characteristics. This result is, however, open to the objection that those workers selected to work with computers were perhaps superior, in the sense that they possessed unobserved attributes which employers' preferred. Thus, Di Nardo and Pischke

(1996) find that the observed correlation between computer use and pay in Germany is indeed capturing unobserved heterogeneity. On the other hand, the longitudinal studies reported in Bell (1996) and Allen (1996) both indicate that there remains a strong relationship between computers or high technology and pay even controlling for both observed and unobserved fixed individual characteristics.

Summary

Is the fall in demand for unskilled workers due to trade or technology? The facts are these.

- There is some evidence that the world price of skill intensive products has risen relative to that of unskilled intensive goods. This is consistent with the trade story but is not inconsistent with the technology argument.
- Looking directly at the skill content of exports and imports reveals that the existence of trade reduces the relative demand for unskilled workers but not by enough to explain the observed relative wage changes.
- Calibrated general equilibrium models provide mixed results but, as yet, there have been no models which satisfactorily capture all the necessary features.
- The fact that there has been an increase in the proportion of skilled workers in establishments in the non-traded sector is inconsistent with a pure trade story.
- There is some persuasive evidence that technological improvements are directly associated with wage premia.

Overall, the balance of the evidence suggests that while LDC trade has had some impact on the demand for the unskilled across the OECD, it is not the major part of the story.

OECD Unemployment: The Background Picture

In order to understand the consequences of the secular decline in demand for unskilled workers in recent decades, it is important to have a clear picture of what unemployment looks like across Europe and North America. In particular, it is vital to avoid the trap of aggregating the different European countries into some all purpose entity called Europe, which can then be compared with the United States. A typical example is provided by Freeman (1995) who notes that "The rise in joblessness in Europe is thus the flip side of the rise in earnings inequality in the U.S. The two outcomes reflect the same phenomenon—a relative decline in the demand against the less skilled that has overwhelmed the long-term decline in the relative supply of less skilled workers. In the United States, where wages are highly flexible, the change in the supply-demand balance lowered the wages of the less skilled. In Europe, where institutions buttress the bottom parts of the wage distribution, the change produced unemployment" (p. 19).

A Picture of Unemployment in Europe and North America

In Table 2, we present a picture of unemployment in OECD Europe as well as North America. We include Canada as well as the United States because, as Card et al. (1995) demonstrate, relative wages there are only slightly less flexible than in the U.S. in contrast to their complete inflexibility in France. So Canada is worth incorporating into the overall comparison.

Taking the period 1983-95,[5] several points are worth making. First, there is enormous variation in European unemployment rates stretching from 1.7 percent in Switzerland to 19.5 percent in Spain. Second, around 30 percent of the population of OECD Europe lives in countries with average unemployment rates lower than the United States and around 70 percent in countries with lower unemployment than Canada. Furthermore, the European countries with the lowest unemployment rates are not noted for their wage flexibility (Austria, West Germany, Norway, Portugal, Sweden, Switzerland). Britain probably has the most flexible labour market but has an average unemployment rate higher than half of Europe.[7] Third, unlike North America, many European countries have exceptionally high rates of long-term unemployment. Thus in Belgium, Ireland, Italy and Spain, more than half the unemployed have durations over 12 months.

Table 2

Unemployment Rates in Europe and North America

	1983–88			1989–94			1995	1983–95
	Total	Short term	Long term	Total	Short term	Long term	Total	Total
Austria	3.6	na	na	3.7	na	na	4.3	3.7
Belgium	11.3	3.3	8.0	8.1	2.9	5.1	9.4	9.7
Denmark	9.0	6.0	3.0	10.8	7.9	3.0	10.6	10.0
Finland	5.1	4.0	1.0	10.5	8.9	1.7	17.0	8.5
France	9.8	5.4	4.4	10.4	6.5	3.9	11.6	10.2
Germany(W)	6.8	3.7	3.1	5.4	3.2	2.2	8.2	6.3
Ireland	16.1	6.9	9.2	14.8	5.4	9.4	12.9	15.3
Italy	6.9	3.1	3.8	8.2	2.9	5.3	8.1	7.6
Netherlands	10.5	5.0	5.5	7.0	3.5	3.5	6.5	8.6
Norway	2.7	2.5	0.2	5.5	4.3	1.2	4.9	4.1
Portugal	7.6	3.5	4.2	5.0	3.0	2.0	7.1	6.4
Spain	19.6	8.3	11.3	18.9	9.1	9.7	22.7	19.5
Sweden	2.6	2.3	0.3	4.4	4.0	0.4	9.2	3.9
Switzerland	0.8	0.7	0.1	2.3	1.8	0.5	4.1	1.7
UK	10.9	5.8	5.1	8.9	5.5	3.4	8.7	9.8
Canada	9.9	9.0	0.9	9.8	8.9	0.9	9.5	9.8
US	7.1	6.4	0.7	6.2	5.6	0.6	5.5	6.6

Notes: These rates are OECD standardised rates[6] with the exception of Italy where we use the U.S. Bureau of Labor Statistics (BLS) "unemployment rates on US concepts".Aside from Italy, the OECD rates and the BLS rates are very similar.For Italy, the OECD rates appear to include the large numbers of Italians who are registered as unemployed but have performed no active job search in the previous 4 weeks.Long-term rates refer to those unemployed with durations over 1 year.The data are taken from the *OECD Employment Outlook* and the UK *Employment Trends*, published by the Dept. of Employment and Education.

One consequence of these numbers is plain. While the notion that relative wage rigidity causes high European unemployment may have some limited value as a short-hand, it can hardly tell the whole story. Unemployment is not high in large parts of Europe where wages are supposedly rigid and it is high in Britain and Canada where relative wage rigidity is not a problem. Furthermore, it is worth remarking that a vast number of potentially unemployed low skill individuals in the United States are in prison and so do not enter the statistics at all. Thus, the number of incarcerated males in the US is equal to about 2 percent of the labour force. More importantly, the number of low-skill males[8] in prison is equal to around 9 percent of the equivalent number in the labour force. This is almost as high as the male low-skill unemployment rate. This has important implications for comparisons with Europe, particularly among low-skill men, because comparatively few of these are in prison in Europe (proportionately, around one eighth of the US numbers). Indeed, Blanchflower and Freeman (1996) go so far as to argue that, for low-skill men, imprisonment is, in some sense, the equivalent in the US of long-term unemployment in Europe, although it is, of course, a lot more expensive.

Some Causes of Unemployment Across the OECD

At this stage, it is useful to get some idea of what generates the enormous variations in unemployment across countries which we see in Table 2.[9] The factors which we consider are the treatment of the unemployed, the system of wage bargaining, employment protection and taxation. We begin by listing some useful variables under these headings and illustrating their impact on unemployment via a regression covering some 20 countries (those in Table 2 plus Japan, Australia, New Zealand). Then we discuss the factors in more detail.

We propose to explain unemployment in two sub-periods, 1983-88 and 1989-94. In Table 3, we list a set of variables for both sub-periods. These are as follows.

- **Treatment of the unemployed:** Benefit replacement ratio (%); Benefit duration (years: indefinite = 4 years); Active labour market policy (ALMP). That is, expenditure per unemployed person as % of output per worker.
- **Wage bargaining system:** Union coverage (1 under 25%, 2 middle, 3 over 75%); Union and employer coordination in wage bargaining (1 low, 2 middle, 3 high).
- **Employment protection:** OECD ranking (1 low, 20 high).

Table 3

Variables Used to Explain Unemployment

When variable changes between the two sub-periods, the first number is for 1983-88 and the second for 1989-94

	Replacement Rate		Benefit Duration		ALMP		Union Coverage	
Belgium	60		4		10.0	14.6	3	
Denmark	90		2.5		10.6	10.3	3	
France	57		3.75	3	7.2	8.8	3	
Germany	63		4		12.9	25.7	3	
Ireland	50	37	4		9.2	9.1	3	
Italy	2	20	0.5		10.1	10.3	3	
Netherlands	70		4		4.0	6.9	3	
Portugal	60	65	0.5	0.8	5.9	18.8	3	
Spain	80	70	3.5		3.2	4.7	3	
UK	36	38	4		7.8	6.4	3	2
Australia	39	36	4		4.1	3.2	3	
New Zealand	38	30	4		15.4	6.8	2	
Canada	60	59	0.5	1	6.3	5.9	2	
USA	50		0.5		3.9	3.0	1	
Japan	60		0.5		5.4	4.3	2	
Austria	60	50	4		8.7	8.3	3	
Finland	75	63	4	2	18.4	16.4	3	
Norway	65		1.5		9.5	14.7	3	
Sweden	80		1.2		59.5	59.3	3	
Switzerland	70		1		23.0	8.2	2	

Replacement rate and benefit duration: Mainly US Department of Health and Social Services, *Social Security Programmes throughout the World,* 1985 and 1993. See Layard et al. (1991), Annex 1.3

ALMP (active labour market policies): *OECD Employment Outlook,* 1988 and 1995. For the first sub-period the data relate to 1987 and for the second to 1991. We include all active spending, except on the disabled.

Union coverage—union coordination and employer coordination: See Layard et al. (1991), Annex 1.4 and OECD *Employment Outlook,* 1994, pp. 175-185. **Employment protection:** OECD Jobs Study (1994) Part II Table 6.7 Col. 5 p. 74. Country ranking with 20 as the most strictly regulated.

Table 3 (Continued)

Union Coordination	Employer Coordination	Employment Protection	Payroll Tax Rate (%)	Total Tax Rate (%)	Change in Inflation
2	2	17	19.6 21.5	47.6 49.8	-0.76-0.52
3	3	5	1.4 0.6	48.8 46.3	-0.86-0.46
2	2	14	38.3 38.8	62.8 63.8	-1.38-0.30
2	3	15	23.2 23.0	52.6 53.0	-0.34-0.04
1	1	12	7.0 7.1	33.6 34.3	-1.52-0.54
2	1 2	20	37.4 40.2	57.2 62.9	-1.68-0.52
2	2	9	30.4 27.5	59.3 56.5	-0.14 0.14
2	2	18	12.6 14.5	33.5 37.6	-2.74-1.28
2	1	19	32.4 33.2	50.1 54.2	-1.24-0.60
1	1	7	15.4 13.8	44.6 40.8	0.16-1.02
2	1	4	2.8 2.5	30.8 28.7	0.02-1.24
2 1	1	2	- -	- -	0.36-1.22
1	1	3	10.8 13.0	37.8 42.7	-0.08-0.84
1	1	1	20.3 20.9	42.6 43.8	-0.04-0.48
2	2	8	15.2 16.5	33.1 36.3	-0.20-0.36
3	3	16	22.3 22.6	54.5 53.7	-0.46 0.06
2 2	3	10	22.5 25.5	59.6 65.9	-0.26-0.72
3	3	11	17.0 17.5	49.9 48.6	-0.34-1.12
3	3	13	36.1 37.8	68.9 70.7	-0.75-1.00
1	3	6	14.5 14.5	40.0 38.6	-0.12-0.50

Payroll tax rate: Defined as the ratio of labour costs to wages (less unity). Note this includes pension payments by employers. Centre for Economic Performance OECD data set.

Total tax rate: Defined as the sum of the payroll tax rate, the income tax rate and the consumption tax rate. The latter are derived from national income and aggregate tax data, so they are average rates. Centre for Economic Performance OECD data set.

Inflation: OECD Economic Outlook.

For further definitions, see text, p. 65.

- **Taxation:** Payroll tax rate (%); Total tax rate (%) = payroll tax rate + income tax rate + excise tax rate.

We also include the change in inflation in the regression because it is always possible to achieve a temporary fall in unemployment by allowing inflation to increase.

In Table 4, we set out the regression results. As well as total unemployment, we also explain long-term and short-term unemployment separately. We do this because it is clear from Table 2 that the variation in short-term unemployment rates across countries is relatively slight compared to the enormous variation in long-term unemployment rates. It appears that countries can live with very different rates of long-term unemployment, whereas some short-term unemployment seems inevitable. The reason for the "optional" nature of long-term unemployment is that it is much less effective in holding back wage inflation than short-term unemployment (OECD 1993, p. 94).

We now consider the role of each of the factors in turn in the light of Table 4 and other results.

Treatment of the Unemployed

Unemployment benefits operate via two mechanisms. First, they reduce the fear of job loss and directly increase wage pressure. Second, they reduce the effectiveness of unemployed people as fillers of vacancies. The effect of benefit levels on unemployment is well documented (see Layard et al. 1991, OECD 1994, Chapter 8), and it appears to have a significant impact at the 10 percent level on total unemployment in table 4. We see from Table 3 that benefit replacement rates in most European countries are rather high with the notable exception of Italy where, until recently, there was effectively no benefit system at all.

As well as being generous, most benefit systems in Europe pay out for very long periods, the notable exceptions here being Portugal, Norway, Sweden and Switzerland.[10] North America also has limited duration benefits and, not surprisingly, all the countries with short benefit durations have low levels of unemployment duration. This again is a well documented relationship (see OECD 1991,

Table 4

Regressions to Explain Log Unemployment Rate
(20 OECD Countries, 1983-8 and 1989-94)

	Total unemployment		Long-term unemployment		Short-term unemployment	
Replacement Rate (%)	0.011	(1.6)	0.004	(0.5)	0.009	(1.2)
Benefit Duration (years)	0.09	(1.3)	0.16	(1.9)	0.04	(0.6)
ALMP (%)	-0.008	(0.7)	-0.03	(2.0)	-0.0008	(0.07)
Union Coverage (1-3)	0.66	(2.7)	0.56	(1.7)	0.54	(2.2)
Coordination (1-3)	-0.68	(3.2)	-0.29	(0.9)	-0.57	(2.4)
Employment Protection (1-20)	-0.005	(0.2)	0.09	(2.7)	-0.04	(1.6)
Change in Inflation (% points p.a.)	-0.17	(1.7)	-0.13	(1.1)	-0.15	(1.6)
Constant	-3.96	(7.3)	-3.28	(2.9)	-3.8	(7.0)
Dummy for 89-94	0.16	(1.9)	0.1	(0.9)	0.16	(2.1)
Log (Short-Term Unemployment)	-	-	0.94	(4.0)	-	-
R^2	0.59		0.81		0.41	
s.e.	0.51		0.59		0.52	
N	40.00		38.00		38.00	

Dependent Variables: the log of
(1) total unemployed as % of labour force,
(2) long-term unemployed (over 1 year) as % of labour force,
(3) short-term unemployed (under 1 year) as % of labour force.
t-statistics in brackets. Estimation uses the Balestra-Nerlove random effects method.

- ALMP is measured by current active labour market spending as % of GDP divided by current unemployment. To handle problems of endogeneity and measurement error this is instrumented by active labour market spending in 1987 as % of GDP divided by average unemployment rate 1977-9.
- The coefficient measure the proportional effect on unemployment of a unit change in an independent variable, where the unit is measured as in Table 3.
- Coordination = 1/2 (Union coordination + Employer coordination).

Chart 7.1.B, for example).

The benefit system is essentially a passive form of expenditure. Some countries also engage in extensive active measures to try and ensure that the unemployed are able and willing to take up employment. These vary in their effectiveness but tend, overall, to reduce long-term unemployment.

Wage Bargaining System

The key features of wage bargaining systems are the extent to which wages are determined by union bargaining (union coverage)[11] and the degree to which employers and unions coordinate their wage determination activities. It appears that union coverage has a significant impact on unemployment (see Table 4), since union activity increases the pressure on wages. However, if employers and unions coordinate their activities, this can offset the deleterious effect of unions. Indeed, the coefficients in table 4 indicate that the maximum level of coordination is just enough to offset the adverse union effect. The story here is that, by coordination, unions and employers can offset the "leapfrogging" behaviour induced by decentralised union wage bargaining.[12] Of course, if union coverage is very low, as in the United States, coordination is neither required nor can it exist because there are no unions which can coordinate.

It is also worth noting here that it is the existence of unions which can introduce rigidity in relative wages since preserving "differentials" may be one of the important aims of at least part of the union sector. However, it is important to bear in mind that a significant offset to this type of rigidity is provided by the existence of a large self-employed sector, where wages are, presumably, totally flexible. In Table 5 we set out the relevant numbers which reveal that in some European countries, particularly in the South, the extent of self-employment is substantial and would have a significant impact on the overall labour market.

Employment Protection

Employment protection refers to the legislative or bargained environment which covers the extent to which firms are restricted

Table 5

Percentage of Self-Employed in Total Employment

	Whole economy	Non-agricultural sector		Whole economy	Non-agricultural sector
Austria	14.2	6.7	Netherlands	11.6	8.1
Belgium	18.1	14.3	Norway	11.8	6.1
Denmark	11.2	6.8	Portugal	30.1	15.9
Finland	14.9	8.8	Spain	27.9	17.9
France	15.0	9.1	Sweden	9.1	7.1
Germany	11.0	8.0	Switzerland	16.9	–
Ireland	24.6	12.8	UK	14.1	12.4
Italy	29.1	22.2	Canada	9.4	7.5
US	10.8	7.7			

Source: OECD (1994), Table 6.8.

in their ability to hire or fire employees. If hiring or firing is difficult or expensive, then firms respond in two ways. First, in downturns, they will reduce their reliance on adjusting the number of employees and increase their dependence on adjusting working hours and work intensity. Second, in upturns, they will be more cautious about hiring because it is expensive to be caught with "too many" employees.

The macroeconomic consequences are that stricter employment protection rules tend to reduce the flow of new entrants into unemployment and to lower the exit probability for those who do become unemployed, thereby increasing their unemployment durations. The impact on the overall unemployment rate is indeterminate. However we can definitely expect a reduction in the inflow and an increase in duration thereby lowering short-term unemployment and raising long-term unemployment. This effect is clear from the results in Table 4 with a negligible impact on the overall unemployment rate.

Taxation

It is commonly argued that payroll taxes have a significant impact on unemployment because they are a "tax on jobs". Generally speaking, this argument is incorrect. Whether employers pay the tax or employees pay the tax will make no difference in a competitive labour market, and also makes little or no difference when wages are determined by union bargaining. All that counts is the tax wedge between labour costs on the one hand and after-tax pay on the other. This explains why countries without payroll taxes, such as Denmark, do not appear to have a superior employment performance. Indeed it makes little difference whether the taxes paid by workers are in the form of income taxes or consumption taxes. Since workers are interested in what their wages will buy, if you cut their income taxes by 10 percent and raise the cost of their consumption by 10 percent, their real wages are unchanged and so are their labour market responses. As a broad generalization, therefore, there is no labour market gain to switching from payroll taxes to consumption taxes.[13]

There is, however, one important exception to this rule. If you increase the pay-roll tax rate by 5 percent and reduce the income tax rate by 5 percent, what happens in equilibrium is that wages fall by 5 percent so that both labour costs and take-home pay are unaffected and, hence, so is employment. But suppose there is a minimum wage and the initial wage happens to be at this level. Then the above story no longer applies. The wage can no longer adjust downwards, labour costs will rise and employment will fall when the pay-roll tax increases. This minimum wage factor may be important in certain circumstances but, overall, there is no general evidence that pay-roll tax rates alone have any significant impact on unemployment.[14]

A related issue is whether the total tax burden on labour has any impact on unemployment. The evidence reported in OECD (1990), Annex 6A indicates that it may have some effect in the short run but not in the long run, because the whole tax burden is shifted onto labour. However if we include the total tax burden in the regression on Table 4 it appears to have a small and almost significant positive impact on unemployment. However, a country would have to reduce its overall tax burden by 6 or 7 percentage points of GDP to reduce unemployment by 1 percentage point, so the effect is hardly dramatic.

Overall Picture

Broadly speaking we have a picture where high unemployment countries are those where unemployment benefits are high, benefit entitlements last for many years, unions are strong, no effort is made to prepare the unemployed for work, wage bargaining is uncoordinated and perhaps the overall tax burden on labour is high. The level of payroll taxes per se and the strictness of employment protection legislation is irrelevant.

In particular individual countries, there are special additional factors at work. For example in France we find very high payroll tax rates (around 38 percent in 1991) plus the highest statutory minimum wages in the OECD (bar the Netherlands) at around 50 percent of average wages. This combination will ensure a significant impact of tax policy on labour costs. In Finland and Sweden, deregulation of the financial system in the late 1980s led to unprecedented credit booms assisted by tax deductibility of interest payments at very high marginal tax rates. This was followed by a collapse in asset prices[15] and consumption, a huge debt overhang and de facto bankruptcy of most of the banks. In Finland, this was reinforced by the collapse of Soviet trade. The consequence of all this has been massive demand led (debt deflation based) slumps in both countries from which neither has yet recovered. The role of labour market rigidities in this has been minimal. We shall return to other examples at a later stage.

Before we go on to discuss the particular consequences of the collapse in demand for the unskilled, it is worth having a brief look at who the unemployed in each country actually are.

Who Are the Unemployed?

In this section, we consider briefly how unemployment is distributed across the working population, focusing particularly on men, women and young people. In Table 6, we present unemployment rates in three categories, husbands, wives and young persons (under 25). In brackets we show the percentage of the unemployed in a household with no wage earner. Single persons over 25 are omitted. The overall picture is one where, with the notable exceptions of Austria, Germany and Switzerland, unemployment rates among the young are uniformly high. Furthermore, in Continental Europe the unemployment rate among husbands is comparatively low. Indeed the unemployment rate among husbands in all the Anglo-Saxon countries (Ireland, UK, US, Canada) is higher than the Continental European average.

The overall picture suggests that the Continental European system is geared to protecting the family by ensuring continuing employment among husbands, while forcing wives to take their chances and young people to queue for suitable jobs while mainly living at home. This pattern is exemplified in its starkest form in Italy, where there are no benefits and unemployed husbands are extremely rare. Employment protection laws are the strictest in the world and this keeps husbands in work while making it hard for new entrants to break into the system, which explains the massive youth unemployment rate. However, once a young man does break into the job market, then he is very secure. It is, however, worth recalling that over one-fifth of total non-agricultural employment is self-employed and so does not enter these statistics.

The Consequences of the Demand Shift Against the Unskilled

Having obtained some idea of the overall unemployment picture in Europe and North America, we may now focus on the fall in demand for unskilled workers. We shall assume that, because both trade and technology shifts have been very similar across all the OECD countries, the demand shift against the unskilled has been a significant factor throughout the OECD. We shall therefore start by setting out information on unemployment and wages by skill for those European and North American countries for which we have good data.

Table 6

Unemployment Rates (%) by Family Status and Percentages in a "No Earner" Household (1991 or 1992)

	Husbands		Wives		Young Persons	
Austria	2.2	(32.9)	3.0	(19.4)	4.1	(21.9)
Belgium	2.5	(63.3)	8.3	(12.7)	13.9	(31.7)
Denmark	5.0		8.4		11.5	
Finland	9.4		8.1		23.5	
France	4.8	(47.6)	10.0	(20.5)	20.8	(29.4)
Germany	2.4	(45.8)	4.9	(22.6)	5.6	(33.8)
Ireland	12.3	(64.4)	16.0	(28.9)	25.3	(38.8)
Italy	2.0	(62.1)	10.5	(16.7)	32.7	(21.3)
Netherlands	3.3	(51.8)	8.5	(14.6)	10.2	(32.0)
Norway	4.0		3.8		13.9	
Portugal	1.3	(32.1)	4.9	(11.4)	9.4	(10.0)
Spain	7.0	(58.9)	18.9	(22.2)	34.4	(24.9)
Switzerland	na		na		4.8	
UK	6.7	(57.9)	5.8	(27.5)	15.2	(33.8)
Canada	8.4		9.1		17.8	
US	4.8	(33.6)	4.8	(17.2)	13.7	(na)
EC	4.3		8.7		18.9	
EFTA	4.2		4.1		9.6	
NA	5.2		5.2		14.1	

The figures in brackets are the percentages in each category living in a household without a wage earner.
• Single adults over 25 do not appear in the table.

Source: OECD (1994), Tables 1.17, 1.19; OECD (1995), Tables 1.12, 1.13.

Unemployment and Wages by Skill

In Table 7, we give the history of male unemployment rates by education levels for a number of countries. We concentrate on male rates because temporal comparisons of male unemployment are easier to interpret, men having had a relatively stable attachment to the labour force over the period. Then in Tables 8, 9, 10 we present a variety of data on wage movements including both wages by education level and the overall wage distribution. The following facts emerge.

- In all countries bar Italy, unemployment among the well educated is far lower than among the low education group. The obvious explanation for the exception of Italy is that, in the absence of a benefit system, the high education group is better able to afford an unemployment spell.
- In all countries we see a large rise in unskilled unemployment from the 1970s to the 1990s. In many countries, we also have a fairly substantial rise in skilled unemployment. However in Germany, Norway, Sweden and the US, the increase in skilled unemployment is relatively slight. Overall, the pattern of the rise in US unemployment from the early 1970s to the 1990s is very similar to that in Germany from the mid 1970s to the 1990s.
- In the 1980s, there has been a very large increase in both earnings dispersion and the education premium in the UK and the US. There has been some increase in dispersion in Canada and some increase in the education premium in Germany. Otherwise, there have been no significant changes in dispersion or education premia in the other European countries.16 This is consistent with the notion that relative wages in the European countries are inflexible, with the notable exception of the UK where they seem highly flexible.

Overall, therefore, we may note that all our countries have seen substantial rises in the unemployment rates of the unskilled whether or not their relative wages are flexible. Some countries have seen a significant rise in the skilled unemployment rate, including inflexible Spain and flexible Britain. Others have had only slight rises in the skilled unemployment rate, including inflexible Germany and flexible America. However, it is worth recalling that some European countries have very high rates of self-employment so that wages may, in fact, be more flexible than would appear from wage data for employees.

More on unemployment and wages

The patterns of unemployment and wages across skill groups do not seem very coherent so, as a next step, we consider three representative groups of countries, high unemployment countries in

Table 7

Male Unemployment Rates by Education (%)

	1971-74	1975-78	1979-82	1983-86	1987-90	1991-93
France						
Total			5.2[a]	6.7[b]	7.2	8.1
High ed.			2.1	2.5	2.6	4.2
Low ed.			6.5	9.0	10.8	12.1
Ratio			3.1	3.6	4.1	2.9
Germany						
Total		2.8	3.4	6.3	4.9	4.1[c]
High ed.		1.5	2.0	3.3	2.9	2.2
Low ed.		5.2	7.6	13.9	12.1	10.7
Ratio		3.5	3.8	4.2	4.2	4.9
Italy						
Total (M+F)		7.2	8.2	10.5	11.8	11.2[c]
High ed.		12.3	12.2	13.1	13.1	12.5
Low ed.		4.4	4.8	6.4	8.1	7.5
Ratio		0.4	0.4	0.5	0.6	0.6
Netherlands						
Total (M+F)		5.5[d]	7.1[e]	13.1[f]	6.9[g]	6.8
High ed.		2.9	3.4	6.2	5.2	5.0
Low ed.		5.7	8.3	18.0	9.9	9.9
Ratio		2.0	2.4	2.9	1.9	2.0
Norway						
Total (M+F)	1.2[h]	1.9	2.1	2.7	3.9	5.7
High ed.	1.0	0.8	0.9	0.8	1.5	2.6
Low ed.	1.9	2.2	2.9	3.8	6.0	8.8
Ratio	1.9	2.8	3.2	4.8	4.0	3.4
Spain						
Total		6.1	11.7	18.5	15.3	15.1
High ed.		4.5	7.9	11.0	8.8	9.0
Low ed.		7.7	13.5	21.4	17.7	20.0
Ratio		1.7	1.7	1.9	2.0	2.2

Table 7 (Continued)

	1971-74	1975-78	1979-82	1983-86	1987-90	1991-93
Sweden						
Total	2.8	1.9	2.4	3.1	1.8	5.8
High ed.	1.3	0.8	0.9	1.1	1.0	2.8
Low ed.	3.2	2.4	3.1	4.1	2.4	6.9
Ratio	2.5	4.0	3.4	3.7	2.4	2.5
UK						
Total	2.9[j]	4.4	7.7	10.5	7.5	10.8[c]
High ed.	1.4	2.0	3.9	4.7	4.0	6.2
Low ed.	4.0	6.4	12.2	18.2	13.5	17.1
Ratio	2.9	3.2	3.1	3.9	3.4	2.8
Canada						
Total		6.9	6.6[k]	10.3[m]	7.8	11.6
High ed.		2.6	2.4	4.3	3.4	5.1
Low ed.		8.2	8.3	12.5	11.3	16.1
Ratio		3.2	3.5	2.9	3.3	3.2
US						
Total	3.6	5.5	5.7	7.3	5.1	6.0
High ed.	1.7	2.2	2.1	2.7	2.1	3.0
Low ed.	5.3	8.6	9.4	12.8	9.8	11.0
Ratio	3.1	3.9	4.5	4.7	4.7	3.7

See notes on page 87

Europe, low unemployment countries in Europe and flexible wage countries. Changes in the unemployment rates of these groups of countries are shown in Table 11.

By the 1990s, the high unemployment countries found themselves with substantial levels of unemployment for skilled workers and extremely high levels for the unskilled. By contrast, the rates in the low unemployment European countries were comparable to or lower than those in the flexible wage countries, despite having wages which are quite as inflexible as their high unemployment European counterparts.[17]

Table 8

Earnings Differentials by Education (Males)

	Ratio of High to Low Education Groups		
	Early 1970s	**Early 1980s**	**Late 1980s**
France		1.66	1.63
Germany		1.36	1.42
Italy	1.96	1.60	1.61
Netherlands		1.50	1.22
Sweden	1.40	1.16	1.19
UK	1.64	1.53	1.65
Canada	1.65	1.40	1.42
U.S.	1.49	1.37	1.51

Source: *OECD Employment Outlook* (1993, Table 5.6); Davis (1992).

So what is going on here? First, it seems clear that the very high level of unemployment in the first group of countries is not specifically due to shifts in relative demand against the unskilled and in favour of the skilled. The big increases in skilled unemployment suggest that some other factors, which are more neutral as between skill levels, must have been important. In France, we have had the imposition of a rigorous monetary policy since the mid 1980s allied to widespread uncoordinated union wage setting. The Spanish labour market has never recovered from the surge in wages and union power following the demise of the Franco regime. Finland was devastated in the early 1990s by the collapse of the enormous domestic credit boom of the late 1980s and the disappearance of Soviet trade. In other words, these high unemployment countries have confronted specific problems which have little to do with relative demand shifts against the unskilled.[18]

The key feature of the low unemployment European countries is the fact they have been able to maintain comparatively low levels of unemployment at the bottom end of the labour market despite the substantial falls in demand for their unskilled labour *and* the apparent inflexibility of their relative wages. Indeed their levels of

Table 9

Earnings Dispersion for Males

	1973	1975	1979-81	1985-86	1987-88	1989-90	1991
France							
D9/D5	2.00	2.09	2.05	2.10*	2.09	2.11	2.11
D1/D5	0.62	0.61	0.63	0.64*	0.66	0.66	0.66
D9/D1	3.23	3.43	3.25	3.28*	3.17	3.20	3.20
Germany							
D9/D5			1.47	1.65*	1.65	1.65	1.65
D1/D5			0.67	0.69*	0.71	0.72	0.71
D9/D1			2.19	2.39*	2.32	2.29	2.32
Italy							
D9/D5			1.44	1.51	1.56		
D1/D5			0.69	0.73	0.75		
D9/D1			2.09	2.07	2.08		
Netherlands							
D9/D5			1.66	1.63		1.65	
D1/D5			0.75	0.75		0.72	
D9/D1			2.21	2.17		2.29	
Norway (M+F)							
D9/D5			1.46		1.49		1.50
D1/D5			0.71		0.69		0.76
D9/D1			2.06		2.16		1.97
Sweden							
D9/D5	1.57		1.68*	1.50	1.56		1.57
D1/D5	0.76		0.78	0.76	0.76		0.73
D9/D1	2.07		2.15	1.97	2.05		2.15
UK							
D9/D5	1.70	1.66	1.72	1.85*	1.91	1.96	1.99
D1/D5	0.68	0.70	0.68	0.63*	0.62	0.61	0.59
D9/D1	2.50	2.37	2.53	2.94*	3.08	3.21	3.37
Canada							
D9/D5	1.67		1.67	1.68	1.71	1.75	
D1/D5	0.52		0.48	0.42	0.45	0.44	
D9/D1	3.21		3.48	4.00	3.80	3.98	
US							
D9/D5		1.93	1.95	2.09	2.10	2.14	
D1/D5		0.41	0.41	0.38	0.38	0.38	
D9/D1		4.71	4.76	5.50	5.53	5.63	

D9, D5, D1 are upper limits of the deciles of the earnings distribution;
* indicates change in measurement, so not comparable to previous numbers.
Source: OECD *Employment Outlook* (1993, Table 5.2).

Table 10

Trends in Real Earnings Deciles for Men

	D1	D5	D9		D1	D5	D9
France				**Sweden**			
1980	100	100	100	1981	100	100	100
1984	103	101	102	1986	110	110	106
1987	103	103	107	1991	121	122	117
Germany				**UK**			
1983	100	100	100	1980	100	100	100
1986	114	111	113	1983	100	106	112
1988	123	119	120	1987	105	116	128
Italy				1991	111	129	151
1980	100	100	100	**Canada**			
1984	109	104	104	1981	100	100	100
1987	114	105	113	1986	87	101	101
Norway(M+F)				1990	92	101	105
1980	100	100	100	**US**			
1983	102	99	101	1980	100	100	100
1987	102	105	108	1984	89	96	101
1991	119	111	114	1989	89	95	104

Gross earnings deflated by the CPI.
Source: OECD (1993), Table 5.3.

---

unskilled unemployment are markedly lower than those in Canada and the UK. How have they managed to achieve this rather surprising outcome?

One hypothesis which is worth pursuing is that the level of skills at the bottom end of the labour market is much higher in the low unemployment European countries than it is in the flexible wage countries. This means that they are better able to cope with a shift in demand in favour of skilled work without having to make substantial wage adjustments. If those at the bottom have the education and training to cope with an up-skilling of their work, the consequent labour market adjustments are relatively slight.

Some evidence on this question is presented in Table 12 and Table 13. The one fact that stands out in the first of these tables is that the level of performance in the lower part of the ability range is far better in the low unemployment European countries and their neighbours (Netherlands, Germany, Sweden, Switzerland) than it is in the flexible wage countries (England, US). This is related to the vastly greater numbers with rigorous vocational qualifications in the former countries than in the latter, as set out in Table 13. It is also worth noting that in a cohort of young persons who completed their compulsory schooling in Germany and the United States in 1978/9, some 80 percent of German youth had obtained either a vocational training certificate or a degree within 12 years. Furthermore all of the remainder bar 1 percent had received some formal post-secondary education or training. By contrast, 31 percent of US school leavers never received any formal training or education after leaving school and 46 percent gained no certificate or degree (see Buechtemann et al. 1993).

In summary, therefore, it may be argued that a significant group of European countries has coped as well, if not better, than the flexible wage countries in the face of the substantial switch in demand away from the unskilled. Their higher levels of education and training in the lower part of the ability range has, so far, enabled them to respond to this switch without the necessity of dramatic wage adjustments and the consequent relative impoverishment of the low wage earners.

Summary and Conclusions

We have studied the causes and consequences of the significant fall in demand for the unskilled across the OECD. Our broad conclusions are as follows.

- The decline in the relative demand for unskilled workers is due more to technological shifts than to globalization, although the latter may have accelerated the process.
- Across the OECD, there are enormous variations in unemployment rates which have little to do with changes in the relative demand for skills. High unemployment may be found in countries where unemployment benefits are high, benefit eligibility lasts for many years, little effort is made to prepare the unemployed for work, unions are strong and

Table 11

Male Unemployment Rates by Skill in Selected Countries

High unemployment Europe

	Finland		France		Spain	
	1987-90	1992	1982	1993	1975-78	1993
Total	4.6	15.5	5.2	9.4	6.1	17.9
High ed.	1.2	6.3	2.1	5.9	4.5	10.7
Low ed.	5.9	18.4	6.5	13.6	7.7	24.0

Low unemployment Europe

	Germany		Netherlands (M+F)		Norway (M+F)	
	1975-78	1992	1975-78	1993	1972-74	1992
Total	2.8	4.3	5.5	7.5	2.1	5.9
High ed	1.5	2.2	2.9	5.4	0.9	2.8
Low ed.	5.2	11.3	5.7	10.9	2.9	8.9

Flexible wage countries

	Canada		UK		US	
	1975-78	1993	1972-74	1992	1971-74	1993
Total	6.9	11.7	4.4	11.5	3.6	6.4
High ed.	2.6	5.3	2.0	6.6	1.7	3.3
Low ed.	8.2	16.6	6.4	16.9	5.3	11.4

The data are from the same sources as those in Table 7.
Finland: Low ed. Basic education only. High ed. Higher education, both lower and upper levels. Males, aged 15-74.
Source: Työvoiman Koulutus ja Ammatit, 1984-1992/3, Statistics Finland.

wage bargaining is uncoordinated, and the overall tax burden on labour is high (although the effect here is small). Payroll taxes, per se, have no impact in the long run and employment protection reduces inflows and outflows but has no overall effect on unemployment.
- Concerning the skills issue, all countries have seen substantial increases in unemployment rates among the unskilled

irrespective of the degree of wage flexibility present in their economies. The high unemployment European countries have also seen large increases in skilled unemployment. Only Britain and the United States have seen large falls in the relative wages of the unskilled.

- A significant group of European countries has coped as well, if not better, than the flexible wage countries in the face of the demand switch away from the unskilled. This is because they provide high levels of education and training to individuals in the lower half of the ability range. This enables them to respond to the demand switch without the apparent necessity of dramatic wage adjustments and the consequent impoverishment of the low wage earners.

Finally, we see here that the role of education and training is not to mitigate the unemployment problem directly but to help maintain wage levels at the bottom end of the labour market. Without adequate education and training, governments are forced to rely more and more heavily on the use of in-work benefits to maintain incomes. While this appears to work quite well in both Britain and the United States, it is clear that it will bring its own problems as more and more people are brought within the system of state handouts.

Table 12

International Test Scores

(a) Distribution of Scores in International Mathematical Tests of 13 year-old pupils. 1963-4 (%)

	France	Netherlands	Germany	England	US
Score (out of 70)					
≤ 5	14	10	8	24	22
6–30	68	57	59	49	62
31–51	16	25	30	22	14
> 51	2	8	3	5	1
Mean score	18	24	25	19	16
cv (sd/mean)	68	67	53	88	82

(b) Scores in International Mathematics Tests for 13 year-old pupils: 1990 (out of 100)

	France	Italy	Switzerland	England	US
Average	64.1	64.0	70.8	59.5	55.3
Top Decile	89.3	88.0	93.3	89.3	82.7
Bottom Decile	37.3	36.5	50.7	32.0	29.3

(c) Percentage of Employees at Various Levels of Document Literacy: 1995

Literacy Level	Netherlands	Germany	Sweden	US
4/5	24.0	22.4	37.9	22.7
3	48.5	41.6	40.6	33.9
2	21.7	30.7	16.7	25.5
1	5.7	5.3	4.8	17.8

Level 1 is the minimal level, level 5 requires a high degree of sophistication.

Sources: (a), (b), Prais (1994); (c) OECD (1995), Table B-1b.

Table 13

Percent of Economically Active Persons with Vocational Qualifications, 1988-91

	France	Netherlands	Germany	Switzerland	Britain
University degrees	7	8	11	11	11
Intermediate vocational qualifications	40	57	63	66	25
Of which					
Technician	7	19	7	9	7
Craft	33	38	56	57	18
No vocational qualifications	53	35	26	23	64

Source: Prais (1994), Table 1.

Notes for Table 7

France. Low ed. no certification or only primary school certificate.High ed. two years' university education or further education college degree or university degree (5.1% of labour force in 1968, 15.8% in 1990).
Source: Enquête sur L'Emploi, INSEE. Data refer to males, age 15+.

(West) Germany. Low ed. no formal qualification (39% of working age population 1976, 28% in 1989).High ed. degree (11.3% of working age population in 1976, 15.9% in 1989).
Source: Buttler and Tessaring (1993), adjusted to be compatible with OECD standardised rate.

Italy. Low ed. lower secondary or less (56% of labour force in 1977, 23.1% in 1992).High ed. upper secondary or higher (18% of labour force in 1977, 35.2% in 1992).
Source: Annuario Statistico Italiano, ISTAT.M+F refers to males and females, age 25-64.

Netherlands. Low ed. basic education or completed junior secondary school or junior vocational education (72.8% of labour force in 1975, 33.1% in 1993).High ed. completed vocational college or university (10.2% of labour force in 1975, 23.9% in 1993).
Source: Dutch Central Bureau of Statistics.M+F refers to males and females, age 15-64.

Norway. Low ed. primary level (64.5% of labour force in 1972, 16.3% in 1993). High ed. university level (9.9% of labour force in 1972, 26.3% in 1993).
Source: Labour Market Statistics, Statistik Sentrallyra. Data refer to men and women, age 16-74.

Spain. Low ed.Illiterate or primary (75.8% of labour force in 1976, 40% in 1993).High ed. Superior (university) 2.6% of labour force in 1976, 5.5% in 1993).
Source: Spanish Labour Force Survey.Refers to males, age 16-64.

Sweden. Low ed. pre-upper secondary school up to 10 year (59.7% of labour force in 1971, 30.6% in 1990).High ed. post-upper secondary education (7.9% of labour force in 1971, 21.7% in 1990).
Source: Swedish Labour Force Surveys.Refers to males, age 16-64.

UK. Low ed.No qualifications (55.7% of labour force in 1973, 28.2% in 1991). High ed.Passed A levels (18+ exam.) or professional qualification or degree (16.4% of labour force in 1973, 36.8% in 1991).
Source: General Household Survey.Refers to males, age 16-64.

Canada. Low ed.Up to level 8 (23.3% of labour force in 1975, 7.3% in 1993). High ed.University degree (10.4% of labour force in 1975, 16.8% in 1993).
Source: The Labour Force, Statistics Canada.Refers to males, age 15+.

US. Low ed.Less than 4 years of high school (37.5% of labour force in 1970, 14.5% in 1991).High ed. 4 or more years of college (15.7% of labour force in 1970, 28.2% in 1991).
Source: Handbook of Labor Statistics, BLS, 1989 (Table 67). *Statistical Abstract of the US* (1993, Table 654).Refers to males, age 25-64.
[a] = 1982 only; [b] = 1983,86; [c] = 1991/2; [d] = 1975,77; [e] = 1979,81; [f] = 1983,85; [g] = 1990; [h] = 1972-4; [j] = 1973-4; [k] = 1979; [m] = 1984-6.

Ratio = low ed. unemployment ÷ high ed. unemployment.

End Notes

[1] See OECD (1994), p. 90.

[2] The Luddites were breaking machines in rural Britain in the early 19TH century. The classic example of machines displacing labour was the destruction of handloom weaving by the power loom in the first half of the 19TH Century in Britain. During the Napoleonic Wars, the handloom weavers were part of the "labour aristocracy" and by 1820 they numbered no less than 240,000. By 1856 there were only 23,000 left.

[3] As well as Sachs and Shatz (1994), we also have Borjas, Freeman and Katz (1992), for example, for the US. For the UK, Gregory and Greenhalgh (1996) generate overall labour demand effects from trade which are not dissimilar to those of Sachs and Shatz, although they do not distinguish between different types of labour.

[4] There is no specialization and constant returns to scale.

[5] For comparison purposes, it is obviously better to take an average over a number of years rather than focus on any particular year, because business cycles are not well coordinated across countries.

[6] It is worth noting that, for some countries, OECD standardised unemployment rates differ quite substantially from domestically produced registered unemployment rates. Thus, in (West) Germany, for example, registered unemployment rates are considerably higher because they include numbers of individuals who are not seeking work and the unemployed are normalised on a measure of the labour force which excludes the self-employed.

[7] This is equally true for the period from 1989, so it is not just a fact associated with Britains' exceptionally high unemployment in the early 1980s.

[8] The low skilled are defined as those who have less than 4 years of high school; around 14.5 percent of the US male labour force in 1991.

[9] Many of the results in this section are taken from Jackman et al. (1996).

[10] The fact that Italy has limited benefit duration is of little consequence since it has practically no benefit.

[11] It is important to distinguish between union membership or density, and union coverage. The former is concerned with the number of employees

who are union members. The latter refers to the extent of union influence in wage determination. These may differ widely. In France, for example, less than 20 percent of workers are union members yet over 90 percent of wages are determined by union bargains.

[12]The externality here is nicely captured by a remark by the one-time British Prime Minister, Harold Wilson, who noted that "One man's wage increase is another man's price increase".

[13]This is not quite correct in the sense that since non-workers as well as workers pay consumption taxes, the consumption tax rate on workers can be that bit lower than the income tax rate while raising the same revenue. This effective lowering of the tax rate on workers may have beneficial effects.

[14]If we add the pay-roll tax rate to the equations in Table 4, it is completely insignificant. See also OECD (1990), Annex 6A.

[15]Real house prices in Finland fell by nearly 50 percent from 1989 to 1993.

[16]Including Spain (see OECD 1993, p. 158).

[17]It is worth commenting on the inclusion of (West) Germany in our list of low unemployment European countries in the light of the current press criticism of the German economy. Several points are worth noting. First, on a comparable basis, West German unemployment is still only around 8.5 percent (OECD standardised rate), which is lower than that in Britain. Of course, unemployment in Eastern Germany is much higher, as it is in nearly all East European transformation economies. Second, gross transfers to the East have been running at around 6 percent of the West's GDP per year. This represents a truly enormous adverse shock, more than comparable to the two oil shocks of the 1970s. The fact that this is having a significant impact on the German economy should come as no surprise.

[18]Which is not to say that the shocks which they have faced have not been exacerbated by the lack of flexibility in their labour markets.

References

Allen, S.G. (1996), "Technology and the Wage Structure", NBER Working Paper 5534, April; Cambridge, Mass.

Bell, B.D. (1996), "Skill-Biased Technical Change and Wages: Evidence from a Longitudinal Data Set", Institute of Economics and Statistics, University of Oxford, mimeo.

Berman, E., Bound, J. and Griliches, Z. (1994), "Changes in the Demand for Skilled Labour within U.S. Manufacturing: Evidence from the Annual Survey of Manufactures", *Quarterly Journal of Economics*, Vol. 109, May, 367-97.

Blanchflower, D.G. and Freeman, R.B. (1996), "Growing into Work", paper presented at the *Conference on Labour Market Changes and Income Dynamics*, March 1996, Centre for Economic Performance, London School of Economics, mimeo.

Borjas, G., Freeman, R., and Katz, L. (1992), "On the Labor Market Effects of Immigration and Trade" in G. Borjas and R. Freeman (eds.) *Immigration and the Work Force* (Chicago: University of Chicago Press and NBER), 213-44.

Buechtemann, C., Schupp, J. and Soloff, D. (1993), "Roads to Work: School-to-Work Transition Patterns in Germany and the US", *Industrial Relations Journal*, Vol. 24, 97-111.

Card, D., Kramarz, F. and Lemieux, T. (1995), "Changes in the Relative Structure of Wages and Employment: A Comparison of the United States, Canada and France", Industrial Relations Section, Working Paper 355, Princeton University, December.

DiNardo, J.E. and Pischke, J-S (1996), "The Returns to Computer Use Revisited: Have Pencils Changed the Wage Structure Too?", M.I.T., mimeo.

Feenstra, R.C. and Hanson, G. (1996), "Foreign Investment, Outsourcing and Relative Wages", NBER Working Paper 5424, January; Cambridge, Mass.

Freeman, R.B. (1995), "Are Your Wages Set in Beijing?", *Journal of Economic Perspectives*, 9(3), Summer, 15-32.

Gregory, M. and Greenhalgh, C. (1996), "International Trade, Deindustrialisation and Labour Demand—An Input-Output Study for the UK 1979-90", Leverhulme Programme on "The Labour Market Consequences of Technical and Structural Change", Discussion Paper No. 1, May, University of Oxford.

Jackman, R., Layard, R. and Nickell, S. (1996), "Combatting Unemployment: Is Flexibility Enough?" (OECD conference paper), Centre for Economic Performance, London School of Economics, mimeo.

Krueger, A. (1995), "Labor Market Shifts and the Price Puzzle Revisited", paper presented at the Conference in honor of Assar Lindbeck entitled "Unemployment and Wage Dispersion: Is There a Tradeoff", Stockholm, June.

Krugman, P. (1994), "Past and Prospective Causes of High Unemployment" in *Reducing Unemployment: Current Issues and Policy Options,* Proceedings of a Symposium in Jackson Hole, Wyoming, sponsored and published by The Federal Reserve Bank of Kansas City.

Lawrence, R.Z. and Slaughter, M.J. (1993), "International Trade and American Wages in the 1980s: Giant Sucking Sound or Small Hiccup", Brookings Papers on Economic Activity (2), 161-226.

Layard, R., Nickell, S., Jackman, R. (1991), *Unemployment: Macroeconomic Performance and the Labour Market* (Oxford: Oxford University Press).

Leamer, E.E. (1995), "A Trade Economists View of US Wages and Globalization", mimeo.

Machin, S. (1996), "Changes in the Relative Demand for Skill in the UK" in A. Booth and D. Snower (eds.) *Acquiring Skills,* Cambridge: CUP.

Minford, P. (1996), "Unemployment in the OECD Countries and its Remedies" in D. Snower and G. de la Dehesa (eds.), *Unemployment Policy: Government Options for the Labour Market* (Cambridge: Cambridge University Press).

OECD (1990), *Employment Outlook* (Paris: OECD).

OECD (1991), *Employment Outlook* (Paris: OECD).

OECD (1993), *Employment Outlook* (Paris: OECD).

OECD (1994), *Jobs Study: Evidence and Explanations* (Paris: OECD).

Prais, S. (1994), "Economic Performance and Education: The Nature of Britain's Deficiencies" in *Proceedings of the British Academy,* Vol. 84, 151-207.

Sachs, J.D. and Shatz, H.J. (1994), "Trade and Jobs in U.S. Manufacturing", *Brookings Papers on Economic Activity* (1), 1-84.

Sachs, J.D. and Shatz, H.J. (1996), "US Trade with Developing Countries and Wage Inequality", *American Economic Review (Papers and Proceedings)* 86(2), 234-39.

Wood, A. (1994), *North-South Trade, Employment and Inequality: Changing Fortunes in a Skill Driven World* (Oxford: Clarendon Press).

Wood, A. (1995), "How Trade Hurt Unskilled Workers", *Journal of Economic Perspectives,* 9(3), Summer, 57-80.

List of Participants

Change and Prosperity:
The Aspen Institute Program on the World Ecomomy
August 17–21, 1996
Aspen, Colorado

Masood Ahmed
Director of the International
  Economics Department
The World Bank
Washington, DC

John Berry
Reporter
The Washington Post
Washington, DC

Gordon Brown, MP
Shadow Chancellor of
  the Exchequer
United Kingdom

John Bussey
Foreign Editor
The Wall Street Journal
New York, NY

E. Gerald Corrigan
Managing Director
Goldman, Sachs & Co.
New York, NY

W. Bowman Cutter
Managing Director
Warburg Pincus
New York, NY

Charles Dallara
Managing Director
Institute of International Finance
Washington, DC

Rudiger Dornbusch, Professor
Massachusetts Institute of
  Technology
Cambridge, MA

William D. Eberle
Chairman
Manchester Associates
Concord, MA

Stanley Fischer
First Deputy Managing Director
International Monetary Fund
Washington, DC

Bill Frenzel
Guest Scholar - Government
The Brookings Institution
Washington, DC

Franciso Garza
International Business Strategist
Mexico

Miranda S. Goeltom
Deputy Assistant Minister
Coordinating Ministry for
Economy, Finance
  and Development Supervision
Indonesia

William Gorham
President
The Urban Institute
Washington, DC

Gerald Greenwald
Chairman and Chief Executive
  Officer
United Airlines
Chicago, IL

Toyoo Gyohten
Senior Advisor
The Bank of Tokyo-Mitsubishi,
  Ltd.
President
Institute for International
  Monetary Affairs
Japan

Robert D. Hormats
Vice Chairman
Goldman Sachs International
New York, NY

Lawrence Katz
Professor of Economics
Department of Economics
Harvard University
Cambridge, MA

Kurt J. Lauk
Member of the Board of
  Management
Mercedes-Benz AG
Germany

Miguel Mancera
Governor
Banco de Mexico
Mexico

Cathy E. Minehan
President and Chief Executive
  Officer
Federal Reserve Bank of Boston
Boston, MA

Yuan Ming
Director
Institute of International Relations
Peking University
People's Republic of China

John V. Moller, Director
Change and Prosperity: The Aspen
  Institute Program on the World
  Economy
c/o Policy Consulting Services, Inc.
Washington, DC

Frank N. Newman
Chairman of the Board and
  Chief Executive Officer
Bankers Trust Company
New York, NY

Stephen Nickell
Institute of Economics and
Statistics
University of Oxford
United Kingdom

Nguyen Xuan Oanh
President
N.X. Oanh Associates, Ltd.
Vietnam

Thomas W. Payzant
Superintendent of Schools
Boston Public Schools
Boston, MA

Rupert Pennant-Rea
Chairman
Caspian Securities Limited
United Kingdom

William R. Rhodes
Vice Chairman
Citibank, N.A.
New York, NY

Agostino Rocca
President and Chief Executive
  Officer
Techint Group
Argentina

Eisuke Sakakibara
Director-General
International Finance Bureau
Ministry of Finance
Japan

Alberto Santos
Presidente Del Consejo De
  Empresas Santos
Mexico

Carlo Scognamiglio
Aspen Institute Italia
Italy

John D. Taylor
Director
Infrastructure, Energy and
  Financial Sectors Department
  (West)
Asian Development Bank
Philippines

Edwin (Ted) M. Truman
Staff Director
Division of International Finance
Federal Reserve System
Washington, DC

Paul Volcker
Chairman
J.D. Wolfensohn Inc.
New York, NY

Daniel Yankelovich
President
Public Agenda
New York, NY

Cornelia Yzer, MP
Member of the German Bundestag
Parliamentary State Secretary
Federal Ministry of Education,
  Science, Research and
  Technology
Germany

Leah J. Zell
Wanger Asset Management, L.P.
Chicago, IL

1996 Conference Agenda

Change and Prosperity:
The Aspen Institute Program on the World Ecomomy

Saturday, August 17

6:30 p.m.   Conference convenes over cocktails/dinner. Spouses and guests are invited.

After-dinner discussion. Several participants will keynote the conference by framing the issues the conference will consider.

Sunday, August 18

9:00 a.m.–   *Macroeconomic Policies for Growth and Adjustment.*
12:30 p.m.

Participants will discuss the state of the world economy and assess the outlook for the foreseeable period ahead, considering what macro-economic policy course corrections the major countries—the G-7 as well as the leading emerging market economy countries—should take to ensure that the global economic expansion endures. We will assess the potential for the major industrial countries to grow at faster rates and consider the implications of the limitations of macro-economic policy in achieving faster rates of growth while simultaneously ensuring price stability and financial stability.

2:00 p.m.– *Integrating Emerging Market Economies Into the Global*
5:30 p.m. *Economy: Public Sector Perspectives on the Challenges Facing Latin America*

We will examine—initially focusing on the issues from the perspective of public sector officials—the extent to which the countries of Latin America have established the essential prerequisites for sustained growth. We will consider whether they have solidified the necessary economic policy fundamentals together with the financial and intellectual (meaning policies and attitudes) infrastructures that are necessary to attract and productively employ the investment capital they need to grow robustly. We will evaluate what concrete actions they need to take to fulfill these prerequisites. We will also take stock of the response of individual countries as well as the international community as a whole to the legacy of the Mexican crisis and the adequacy of measures so far put into place to anticipate and prevent such crises in the future and manage them more effectively if they occur.

DINNER

We will divide into three groups and dine at restaurants in the town of Aspen. Spouses and guests are, of course, welcome.

Monday, August 19

9:00 a.m.– *Integrating Emerging Market Economies Into the Global*
12:00 p.m. *Economy: Private Sector Perspectives on the Challenges Facing Latin America*

We will resume our discussion of the issues addressed in our prior session but focus in this session from the vantage point of the private sector, both industrial and financial. We will evaluate, taking into account perspectives from both from inside and outside Latin

America, the practical experiences of private sector representatives in attracting investment funds on global private capital markets and employing them productively. We will consider what these experiences tell us about the markets' assessment of the adequacy of the policy fundamentals and the financial and intellectual infrastructures Latin American countries have established. We will also consider what this assessment may imply in terms of the need for additional policy reforms.

12:30 p.m.– *Presentation by Daniel Yankelovich on the Socio-*
2:00 p.m.   *economics of Growth and Income Distribution*

Note: we will gather together for lunch in a private room at the Aspen Meadows dining facility on campus. Mr. Yankelovich will speak after lunch. Spouses and guests are welcome to attend.

REMAINDER OF AFTERNOON AND EVENING FREE.

Tuesday, August 20

9:00 a.m.– *The Socio-economics of Growth and Income Distribution:*
12:00 p.m. *Diagnosing U.S. and European Employment Maladies*

We will evaluate the validity of the prevailing conventional wisdom diagnosing the origins of the serious employment problems afflicting the United States and the European Union—stagnating real incomes and widening income disparities in the United States and structural unemployment in Europe. We will examine in particular the possible impact upon the major industrial countries of the greater integration of emerging market economies into the global economic and financial system and consider the implications for both the industrial and emerging market economy countries.

1:30 p.m.–   *The Socio-economics of Growth and Income Distribution:*
4:30 p.m.   *Prescriptions for Progress*

>Building upon our prior discussion, we will consider what specific, concrete steps the United States should take to remedy wage stagnation and growing income inequality and the steps Europe should take to address structural unemployment that together threaten to inhibit the process of global economic integration.

>**DINNER**
>
>We will dine at Toklat Lodge, a rustic setting for a campfire dinner several miles outside Aspen. We will depart Aspen Meadows together as a group at about 5:00 p.m. Spouses and guests are welcome.

Wednesday, August 21

9:00 a.m.–   *The Politics of Economic Policymaking*
12:30 p.m.

>To conclude, we will reflect on the politics of economic policy making in both industrial and emerging market economies. We will focus upon the critical question of how to ensure that democratic governments embrace sensible policies in the face of short term political pressures to act in ways that are inconsistent with their nations' long term needs.
>
>We will consider how governments can build and sustain the political will that is essential to: (a) maintain the fiscal and monetary discipline that is integral to sound economic policy fundamentals; and (b) create and maintain truly open markets for goods, services, and capital that defines full integration with the global economy.

Questions for Discussion

Change and Prosperity:
The Aspen Institute Program on the World Economy
August 17–21, 1996
Aspen, Colorado

SESSION I   Sunday, August 18, 1996   9:00 a.m.–12:30 p.m.
Macroeconomic Policies for Growth and Adjustment

1. How should we assess the overall state of the world economy? Worldwide growth is expected to average approximately 3.8 percent this year, slightly higher than in 1995. The world is generally free of economic crises. The growth of trade has slackened a bit but is nevertheless expanding at a healthy pace. Inflation is down virtually everywhere. Budget deficits and interest rates have declined in many countries. The picture is not uniformly rosy, of course. Low growth and high unemployment plague Europe. Japan's recovery from its worst post-War economic crisis remains fragile. Emerging market countries, despite their robust growth rates overall, confront a broad array of challenges. What is the outlook for the period ahead?

2. What are the main obstacles now facing industrial and developing countries in their quest to maintain robust levels of non-inflationary growth and rising productivity, incomes, and employment levels? How can these obstacles be overcome? What macroeconomic policy corrections are required in the major countries to establish a solid foundation to ensure a sustainable, non-inflationary economic expansion?

3. What are the limits of macroeconomic policy in achieving dynamic rates of growth overall while promoting simultane-

ously price stability and the safety and stability of the financial systems whose role is to mobilize and allocate savings efficiently? What other policy tools need to be brought to bear to meet these challenges?

4. At Halifax in 1995, the G-7 endorsed a multi-faceted strategy to improve the response of the international financial system, especially the IMF, to the implications of the globalization of financial markets. This strategy included financial data disclosure standards, an expansion of funds available to the IMF in financial crises (the General Agreements to Borrow), a call for recommendations on handling future liquidity crises of sovereign borrowers, and ways to enhance cooperation in the supervision of financial institutions among others. Where do these initiatives stand? What still needs to be done? Are the additional objectives that were set by the G-7 this year in Lyon with respect to cooperation among financial regulatory and supervisory authorities adequate to the challenges we are likely to face?

SESSION II   Sunday, August 18, 1996   2:00 p.m.–5:30 p.m.

Integrating Emerging Market Economies into the Global–Economy: Public Sector Perspectives on the Challenges Facing Latin America

1. What is the status and outlook for the economies of Latin America? (Can valid generalizations be made for the region as a whole or does the diversity of countries defy reaching broad conclusions?) To what extent have countries put into place the essential prerequisites for achieving sustained economic growth? Are investment and savings rates sufficient? Which countries appear to be on a path to success? Which are at risk? What accounts for the differences? What are the most important policy adjustments required of those at risk?

2. Twenty months after Mexico's liquidity crisis struck, what is the status of Mexico's battle to recover? What is the outlook? What does Mexico's return to private capital markets tell us? What hurdles does Mexico yet face on its path to full recovery? What lessons has Mexico learned from its experience and what are the implications for future policy adjustments and reforms?

3. In retrospect, how great a shock did the "tequila effect" of Mexico's crisis administer to Mexico's Latin neighbors and how great a risk did it actually pose to them? Was the turmoil in global financial markets a rational response to what occurred in Mexico or an overreaction? What is the legacy today in terms of private capital flows to the region? Can and should Latin American countries influence the magnitude and composition of potentially volatile capital inflows? Have Latin American countries built the financial (i.e., institutional) and intellectual (i.e., policies and attitudes) infrastructures necessary to attract and productively employ the investment capital they require to achieve robust rates of growth? What specific actions should they be taking to strengthen those infrastructures?

4. From the perspective of Latin America, do the recommendations of the G-10 Deputies—"The Resolution of Sovereign Liquidity Crises"—constitute an appropriate and sufficient response to improving the environment for, as well as the management of, such crises should they occur in the future? If not, what more needs to be done?

SESSION III Monday, August 19, 1996   9:00 a.m.–12:30 p.m.
Integrating Emerging Market Economies into the Global Economy: Private Sector Perspectives on the Challenges Facing Latin America

1. What has been the experience of private sector representatives, both industrial and financial and from both inside and outside the region, in attracting investment funds from global private capital markets and employing those funds productively? What do these experiences tell us about the markets' assessment of the adequacy of the policy fundamentals and the financial and intellectual infrastructures that Latin American countries have developed? What do they imply in terms of the need for additional policy reforms?

2. Gross capital flows into Latin American countries have grown from under $10 billion per year in the mid-1980s to more than $80 billion in 1994. Average annual net capital flows soared from a net outflow of $17 billion from 1983 to 1989 to net inflows of $40 billion from 1990-1994. Equity has become the

main cross-border financing vehicle for emerging markets worldwide and Latin America is no exception to this trend. Foreign portfolio investment in Latin America was virtually non-existent before 1982 but has accounted for almost a third of capital flows into the region since. Have Latin American equity markets reached their full potential as engines of growth for the region? If not, what needs to be done to enable them to achieve that potential? Is deeper integration with global capital markets necessary? What are the prerequisites for this?

3. Are the pace and direction of structural reform—specifically privatization and deregulation—in Latin America adequate to accelerate growth rates? What has been accomplished to date? What remains to be done? What obstacles lie in the way? How can they be removed?

4. How important a prerequisite to establish conditions for sustained economic growth, including greater foreign investment, in Latin America is labor market reform? What should be done?

5. In what direction, on what terms, and at what pace should Latin America integrate with other regions? Where are NAFTA and MERCUSOR headed? Where are MERCUSOR and the European Union headed? What are the implications for greater integration between North and South America of the United States' inability to reach a political consensus between Congress and the White House on the appropriate terms of future trade liberalization initiatives? What are the implications for the basic domestic policy directions likely to be taken by Latin American countries of closer integration with Europe vs. the United States?

Session IV Tuesday, August 20, 1996 9:00 a.m.–12:00 p.m.

The Socio-economics of Growth and Income Distribution: Diagnosing U.S. and European Employment Maladies

1. What are the appropriate diagnoses for the employment maladies afflicting the United States and Europe? How should those maladies be defined? To what extent do they reflect a mis-

match between the demand for skills required by economies that are constantly transforming themselves due to the inexorable march of technological change and the skills possessed by workers in these countries? To what extent is the problem a reflection of increased imports of manufactured goods from low-wage countries, in other words from the impact upon the industrial countries of the greater integration of emerging market economies into the global economy? To what extent are still other factors responsible? How can the similarities and differences in the experiences of industrial countries best be understood? What implications should we draw?

2. To what extent do the employment problems of the U.S. and Europe reflect a failure of traditional demand management, to what extent are they structural in nature, and to what extent do they reflect the legacy of bad economic and social policy choices of the past? How do we explain the fact that long-term growth rates in the industrial countries have been cut in half—from about 5 percent to about 2.5 percent—since the early 1970s? What are the relationships among changes in rates of growth, levels of employment, and real incomes?

3. Are the goals of ever greater efficiency and higher productivity, a more open international trading system, global financial deregulation, and high wage jobs at "full employment" compatible? Do we face an inevitable choice between creating high levels of employment at the cost of stagnant real wages and growing income inequality on the one hand and creating high wages and high levels of social protection at the cost of rising unemployment on the other?

SESSION V   Tuesday, August 20, 1996   1:30 p.m.–5:00 p.m.
The Socio-economics of Growth and Income Distribution: Prescriptions for Progress

1. What would constitute an appropriate menu of policies to achieve higher rates of growth, higher levels of employment and rising living standards without expanding the already too-high levels of public indebtedness? Where are the largest gains

likely to be reaped? What are the appropriate roles for the public and private sectors?

2. What specific steps should the United States and the countries of Europe take to address the employment problems confronting them? What should their respective priorities be?

3. How great a threat to the process of global economic integration do these problems that we have been discussing pose? What are the implications of our failure to solve them?

SESSION VI Wednesday, August 21, 1996   9:00 a.m.–12:30 p.m.

The Politics of Economic Policy Making

1. What are the basic dynamics of the politics of economic policy making? Are there fundamental, or even significant, differences between this process in industrial and emerging market economy countries?

2. What can be done to ensure that democratic governments in both industrial and emerging market economy countries embrace sensible policies in the face of short-term political pressures to act in ways that are inconsistent with their nations' long-term needs?

3. How can governments build and sustain the political will to: (a) maintain the fiscal and monetary discipline that is integral to sound economic policy fundamentals; and (b) create and maintain open markets for goods, services, and capital to facilitate full integration with the global economy?

4. Are there new roles and responsibilities the G-7—or perhaps a reconstituted G-"X" with appropriate representation from among emerging market economy countries—and existing international financial institutions, specifically the IMF, the World Bank, and the WTO, should assume to bolster the necessary political will in countries to encourage them to adopt and persevere in pursuing sound economic policies?